Solar Electricity Handbook

A simple, practical guide to solar energy: how to design and install photovoltaic solar electric systems

2015 Edition
Internet Linked

Michael Boxwell

Greenstream Publishing

Greenstream Publishing
4 Hargrave Close
Binley
Coventry
CV3 2XS
United Kingdom

www.GreenstreamPublishing.com

Published by Greenstream Publishing 2015

Printed Edition: ISBN 978-1-907670-45-9

First Edition – published April 2009
Second Edition – published November 2009
Third Edition – published March 2010
Fourth Edition – published January 2011
Fifth Edition – published October 2011
Sixth Edition – published February 2012
Seventh Edition – published December 2012
Eigth Edition – published December 2013
Ninth Edition – published January 2015

Michael Boxwell asserts the moral right to be identified as the author of this work.

A catalogue record for this book is available from the British Library.

Table of Contents

Introducing Solar Energy

Ninety-three million miles from Earth, our sun is 333,000 times the size of our planet. It has a diameter of 865,000 miles, a surface temperature of 5,600°C and a core temperature of 15,000,000°C. It is a huge mass of constant nuclear activity.

Directly or indirectly, our sun provides all the power we need to exist and supports all life forms. The sun drives our climate and our weather. It's a huge energy source. Without it, our world would be a frozen wasteland of ice-covered rock.

So harnessing that power from the sun and using it to power electrical equipment is a terrific idea. There are no ongoing electricity bills, no reliance on a power socket: a free and everlasting source of energy that does not harm the planet!

Of course, the reality is a little different from that. Yet generating electricity from sunlight alone is a powerful resource, with applications and benefits throughout the world.

But how does it work? For what is it suitable? What are the limitations? How much does it cost? How do you install it? This book answers all these questions and shows you how to use the power of the sun to generate electricity yourself.

Along the way, I will also expose a few myths about some of the wilder claims made about solar energy and I will show you where solar power may only be part of the solution. Although undoubtedly there are some significant environmental benefits of solar electricity, I will also be talking about where its environmental credentials have been oversold.

I will keep the descriptions as straightforward as possible. There is some mathematics and science involved. This is essential to allow you to plan a solar electric installation successfully. However, none of it is complicated and there are plenty of short-cuts to keep things simple.

The book includes a number of example projects to show how you can use solar electricity. Some of these are very straightforward, such as providing electrical light for a shed or garage, for example, or fitting a solar panel to the roof of a caravan or boat. Others are more complicated, such as installing photovoltaic solar panels to a house.

I also show some rather more unusual examples, such as the possibilities for solar electric motorbikes and cars. These are examples of what can be achieved using solar power alone, along with a little ingenuity and determination.

I have used one main example throughout the book: providing solar-generated electricity for a holiday home which does not have access to an electricity supply from the grid. I have created this example to show the issues and pitfalls that you may encounter along the way, based on real life issues and practical experience.

A website accompanies this book. It has lots of useful information, along with lists of suppliers and a suite of online solar energy calculators that will simplify the cost analysis and design processes.

The website is at *www.SolarElectricityHandbook.com*.

Who this book is aimed at

If you simply want to gain an understanding about how solar electricity works then this handbook will provide you with everything you need to know.

If you are planning to install your own stand-alone solar power system, this handbook is a comprehensive source of information that will help you understand solar and guide you in the design and installation of your own solar electric system.

Solar has a big application for integrating into electrical products: mobile phones, laptop computers, portable radios. Even light electric cars can use solar energy to provide some or all of their power requirements, depending on the application. If you are a designer, looking to see how you can integrate solar into your product, this book will give you a grounding in the technology that you will need to get you started.

If you are specifically looking to install a grid-tie system, i.e. a solar energy system that will feed electricity back into your local power grid, this book will provide you with a good foundation and will allow you to carry out the design of your system. You will still need to check the local planning laws and any other local legislation surrounding the installation of solar energy systems, and you will have to understand the building of electrical systems. In some countries, you specifically need to be certified in order to carry out the physical installation of a grid-tie system.

If you are planning to install larger, commercial–size systems, or if you are hoping to install grid-tie solar systems professionally, then this book will serve as a good introduction, but you will need to grow your knowledge further. This book gives you the foundations you need in order to build this knowledge, but there are special skills required when designing and implementing larger scale solar systems that go far beyond what is required for smaller systems and are beyond the scope of this book.

If you are planning your own solar installation, it will help if you have some DIY skills. Whilst I include a chapter that explains the basics of electricity, a familiarity with wiring is also of benefit for smaller projects and you will require a thorough understanding of electrical systems if you are planning a larger project such as powering a house with solar.

The rapidly changing world of solar energy

I wrote the first edition of this book early in 2009. This 2015 issue is the ninth edition. Most editions have included significant rewrites in order to keep up with the rapid pace of change.

The rapid improvement in the technology and the freefall in costs since early 2009 have transformed the industry. Systems that were completely unaffordable or impractical just two or three years ago are now cost-effective and achievable.

Solar panels available today are smaller, more robust and better value for money than ever before. For many more applications, solar is now the most cost-effective way to generate electricity.

Over the coming years, all the signs are that the technology and the industry will continue to evolve at a similar pace. In the 2012 edition I claimed that by 2015, solar will be the cheapest form of electricity generator, undercutting traditionally low-cost electricity generators such as coal-fired power stations. I was wrong. In reality, we reached that cross-over point in many parts of the world by the middle of 2013 and huge solar farms are becoming a common sight – not just in wealthy parts of the world with hot, sunny climates such as in California and the southern parts of Europe, but in Canada and northern Europe where weather is less predictable; and in India and China, for whom solar was an unaffordable luxury only two or three years ago.

Between 2009 and early 2014, prices of solar panels were in freefall, dropping by as much as one-third each year. By the middle of 2014 prices began to level out. As a consequence, the cost of a solar panel today is around one fifth of its price six years ago.

As prices have fallen and efficiencies improved, we have seen solar energy incorporated into more everyday objects such as laptop computers, mobile phones, backpacks and clothing. Meanwhile, solar energy is causing a revolution for large areas of Asia and Africa, where entire communities are now gaining access to electricity for the first time.

As an easy-to-use and low-carbon energy generator, solar is without equal. Its potential for changing the way we think about energy in the future is huge. For families and businesses in rural African and Asian villages, it is creating a revolution.

Solar electricity and solar heating

Solar electricity is produced from sunlight shining on photovoltaic solar panels. This is different to solar hot water or solar heating systems, where the power of the sun is used to heat water or air.

Solar heating systems are beyond the remit of this book. That said, there is some useful information on surveying and positioning your solar panels later on that is relevant to both solar photovoltaics and solar heating systems.

If you are planning to use solar power to generate heat, solar heating systems are far more efficient than solar electricity, requiring far smaller panels to generate the same amount of energy.

Solar electricity is often referred to as photovoltaic solar, or PV solar. This describes the way that electricity is generated in a solar panel.

For the purposes of this book, whenever I refer to *solar panels* I am talking about photovoltaic solar panels for generating electricity, and not solar heating systems.

The source of solar power

Deep in the centre of the sun, intense nuclear activity generates huge amounts of radiation. In turn, this radiation generates light energy called photons. These photons have no physical mass of their own, but carry huge amounts of energy and momentum.

Different photons carry different wavelengths of light. Some photons will carry non-visible light (*infra-red* and *ultra-violet*), whilst others will carry visible light (*white light*).

Over time, these photons push out from the centre of the sun. It can take one million years for a photon to push out to the surface from the core. Once they reach the sun's surface, these photons rush through space at a speed of 670 million miles per hour. They reach earth in around eight minutes.

On their journey from the sun to earth, photons can collide with and be deflected by other particles, and are destroyed on contact with anything that can absorb radiation, generating heat. That is why you feel warm on a sunny day: your body is absorbing photons from the sun.

Our atmosphere absorbs many of these photons before they reach the surface of the earth. That is one of the two reasons that the sun feels so much hotter in the middle of the day. The sun is overhead and the photons have to travel through a thinner layer of atmosphere to reach us, compared to the end of the day when the sun is setting and the photons have to travel through a much thicker layer of atmosphere.

This is also one of the two reasons why a sunny day in winter is so much colder than a sunny day in summer. In winter, when your location on the earth is tilted away from the sun, the photons have to travel through a much thicker layer of atmosphere to reach us.

Diagram A
Midday

Diagram B
Evening

(The other reason that the sun is hotter during the middle of the day than at the end is because the intensity of photons is much higher at midday. When the sun is low in the sky, these photons are spread over a greater distance simply by the angle of your location on earth relative to the sun.)

The principles of solar electricity

A solar panel generates electricity using the *photovoltaic effect*, a phenomenon discovered in the early 19th century when scientists observed that certain materials produced an electric current when exposed to light.

Two layers of a semi-conducting material are combined to create this effect. One layer has to have a depleted number of electrons. When exposed to sunlight, the layers of material absorb the photons. This excites the electrons, causing some of them to 'jump' from one layer to the other, generating an electrical charge.

The semi-conducting material used to build a solar cell is silicon, cut into very thin wafers. Some of these wafers are then 'doped' to contaminate them, thereby creating an electron imbalance in the wafers. The wafers are then aligned together to make a solar cell. Conductive metal strips attached to the cells take the electrical current.

When a photon hit the solar cell, it can do one of three things: it can be absorbed by the cell, reflected off the cell or pass straight through the cell.

It is when a photon is absorbed by the silicon that an electrical current is generated. The more photons (i.e. the greater intensity of light) that are absorbed by the solar cell, the greater the current generated.

Solar cells generate most of their electricity from direct sunlight. However, they also generate electricity on cloudy days and some systems can even generate very small amounts of electricity on bright moonlit nights.

Individual solar cells typically only generate tiny amounts of electrical energy. To make useful amounts of electricity, these cells are connected together to make a solar module, otherwise known as a solar panel or, to be more precise, a photovoltaic module.

Understanding the terminology

In this book, I use various terms such as 'solar electricity', 'solar energy' and 'solar power'. Here is what I mean when I am talking about these terms:

Solar power is a general term for generating power, whether heat or electricity, from the power of the sun.

Solar energy refers to the energy generated from solar power, whether electrical or as heat.

Solar electricity refers to generating electrical power using photovoltaic solar panels.

Solar heating refers to generating hot water or warm air using solar heating panels or ground-source heat pumps.

Setting expectations for solar electricity

Solar power is a useful way of generating modest amounts of electricity, so long as there is a good amount of sunlight available and your location is free from obstacles such as trees and other buildings that will shade the solar panel from the sun.

In the past, solar experts would have told you that solar electricity is normally only cost-effective where there is no other source of electricity available. That is no longer always the case, but it is certainly true that you need to take a good hard look at the electricity you consume and see how you can keep demand to a minimum. As consumers, it is very easy to underestimate how much electricity we use and solar power can end up becoming very expensive if you simply want to match your typical energy consumption without looking to make savings.

Of course, it is possible to put in a cheap and simple solar energy system, which doesn't cost a fortune and can give you decades of reliable service. Some examples include:

- Installing a light or a power source somewhere where it is tricky to get a standard electricity supply, such as in the garden, shed or remote garage
- Creating a reliable and continuous power source where the standard electricity supply is unreliable because of regular power cuts

- Building a mobile power source that you can take with you, such as a power source for use whilst camping, working on outdoor DIY projects or working on a building site
- Creating green energy for your own use and selling surplus energy production back to the electricity suppliers through a *feed-in tariff*

The amount of energy you need to generate has a direct bearing on the size and cost of a solar electric system. The more electricity you need, the more difficult and more expensive your system will become.

If your requirements for solar electricity are to run a few lights, to run some relatively low-power electrical equipment such as a laptop computer, a small TV, a compact fridge and a few other small bits and pieces, then if you have a suitable location you can achieve what you want with solar.

On the other hand, if you want to run high-power equipment such as fan heaters, washing machines and power tools, you are likely to find that the costs will rapidly get out of control.

As I mentioned earlier, solar electricity is not well suited to generating heat: heating rooms, cooking and heating water all take up significant amounts of energy. Using electricity to generate this heat is extremely inefficient. Instead of using solar electricity to generate heat, you should consider a solar hot water heating system, and heating and cooking with gas or solid fuels.

It is possible to power the average family home purely on solar electricity without making any cuts in your current electricity consumption. However, it is not cheap, and you will need a lot of roof space to fit all the panels! It is usually a good idea to carefully evaluate your electricity usage and make savings where you can before you proceed.

Most households and businesses are very inefficient with their electrical usage. Spending some time first identifying where electricity is wasted and eliminating this waste is an absolute necessity if you want to implement solar electricity cost-effectively.

This is especially true if you live in cooler climates, such as Northern Europe or Canada, where the winter months produce much lower levels of solar energy. In the United Kingdom, for instance, the roof of the average-sized home is not large enough to hold all the solar panels that would be required to provide the electricity used by the average household throughout the year. In this instance, making energy savings is essential.

For other applications, a solar electric installation is much more cost-effective. For instance, no matter which country you live in, providing electricity for a holiday home is well within the capabilities of a solar electric system, so long as heating and cooking are catered for using gas or solid fuels and the site is in a sunny position with little or no shade. In this scenario, a solar electric system may be more cost-effective than installing a conventional electricity supply if the house is *off-grid* and is not close to a grid electricity connection.

If your requirements are more modest, such as providing light for a lock-up garage, for example, there are off-the-shelf packages to do this for a very reasonable cost. Around £50–£80 ($80–$130) will provide you with a lighting system for a shed or small garage, whilst £175 ($280) will provide you with a system big enough for lighting large stables or a workshop.

This is far cheaper than installing a conventional electricity supply into a building, which can be expensive even when a local supply is available just outside the door.

Low-cost solar panels are also ideal for charging up batteries in caravans and recreational vehicles or on boats, ensuring that the batteries get a trickle charge between trips and keeping the batteries in tip-top condition whilst the caravan or boat is not in use.

Why choose a solar electric system?

There are a number of reasons to consider installing a solar electric system:

- Where there is no other source of electrical power available, or where the cost of installing conventional electrical power is too high
- Where other sources of electrical power are not reliable. For example, when power cuts are an issue and a solar system can act as a cost-effective contingency
- When a solar electric system is the most convenient and safest option. For example, installing low voltage solar lighting in a garden or providing courtesy lighting in a remote location
- You can become entirely self sufficient with your own electrical power
- Once installed, solar power provides virtually free power without damaging the environment

Cost-justifying solar

Calculating the true cost of installing a solar electric system depends on various factors:

- The power of the sun at your location at different times of the year
- How much energy you need to generate
- How good your site is for capturing sunlight

Compared to other power sources, solar electric systems typically have a comparatively high capital cost, but a low ongoing maintenance cost.

To create a comparison with alternative power sources, you will often need to calculate a payback of costs over a period of a few years in order to justify the initial cost of a solar electric system.

On all but the simplest of installations, you will need to carry out a survey on your site and carry out some of the design work before you can ascertain the total cost of installing a photovoltaic system. Do not panic: this is not as frightening as it sounds. It is not difficult and I cover it in detail in later chapters.

We can then use this figure to put together a cost-justification on your project to compare with the alternatives.

Solar power and wind power

Wind turbines can be a good alternative to solar power, but probably achieve their best when implemented *together* with a solar system: a small wind turbine can generate electricity in a breeze even when the sun is not shining.

Small wind turbines do have some disadvantages. Firstly, they are very site-specific, requiring higher than average wind speeds and minimal turbulence. They must be higher than surrounding buildings and away from tall trees. If you live on a windswept farm or close to the coast, a wind turbine can work well. If you live in a built-up area or close to trees or main roads, you will find a wind turbine unsuitable for your needs.

Compared to the large wind turbines used by the power companies, small wind turbines are not particularly efficient. If you are planning to install a small wind turbine in combination with a solar electric system, a smaller wind turbine that generates a few watts of power at lower wind speeds is usually better than a large wind turbine that generates lots of power at high wind speeds.

Fuel cells

Fuel cells can be a good way to supplement solar energy, especially for solar electric projects that require additional power in the winter months, when solar energy is at a premium.

A fuel cell works like a generator. It uses a fuel mixture such as methanol, hydrogen or zinc to create electricity.

Unlike a generator, a fuel cell creates energy through chemical reactions rather than through burning fuel in a mechanical engine. These chemical reactions are far more carbon-efficient than a generator.

Fuel cells are extremely quiet, although rarely completely silent, and produce water as their only emission. This makes them suitable for indoor use with little or no ventilation.

Grid-tied solar electric systems

Grid-tied solar electric systems connect directly into the electricity grid. When the sun is shining during the day, excess electricity feeds into the grid. During the evening and night, when the solar panels are not providing sufficient power, electricity is taken from the grid as required.

Grid-tied solar electric systems effectively create a micro power station. Electricity can be used by other people as well as yourself. In some countries, owners of grid-tied solar electric systems receive payment for each kilowatt of power they sell to the electricity providers.

Because a grid-tied solar electric system becomes part of the utility grid, the system will switch off in the event of a power cut. It does this to stop any current flowing back into the grid, which could be fatal for engineers repairing a fault.

Solar electricity and the environment

Once installed, a solar electric system is a low-carbon electricity generator: the sunlight is free and the system maintenance is extremely low.

There is a carbon footprint associated with the manufacture of solar panels, and in the past this footprint has been quite high, mainly due to the relatively small volumes of panels being manufactured and the chemicals required for the 'doping' of the silicon in the panels.

Thanks to improved manufacturing techniques and higher volumes, the carbon footprint of solar panels is now much lower. You can typically offset the carbon footprint of building the solar panels by the energy generated within 2–5 years, and some of the very latest amorphous thin-film solar panels can recoup their carbon footprint in as little as six months.

Therefore, a solar electric system that runs as a complete stand-alone system can reduce your carbon footprint, compared to taking the same power from the grid.

Grid-tied solar systems are slightly different in their environmental benefit, and their environmental payback varies quite dramatically from region to region, depending on a number of factors:

- How grid electricity is generated by the power companies in your area (coal, gas, nuclear, hydro, wind or solar)
- Whether or not your electricity generation coincides with the peak electricity demand in your area (such as air conditioning usage in hot climates, or high electrical usage by nearby heavy industry)

It is therefore much more difficult to put an accurate environmental payback figure on grid-tied solar systems.

It is undeniably true that some people who have grid-tied solar power actually make no difference to the carbon footprint of their home. In colder climates, the majority of electricity consumption is in the evenings and during the winter. If you have grid-tie solar but sell most of your energy to the utility companies during the day in the summer and then buy it back to consume in the evenings and in the winter, you are making little or no difference to the overall carbon footprint of your home. In effect, you are selling your electricity when there is a surplus and buying it back when there is high demand and all the power stations are working at full load.

In warmer climates, solar energy can make a difference. In a hot area, peak energy consumption tends to occur on sunny days as people try to keep cool with air conditioning. In this scenario, peak electricity demand occurs at the same time as peak energy production from a solar array, and a grid-tie solar system can be a perfect fit.

If you live in a colder climate, this does not mean that there is no point in installing a grid-tie solar system. It does mean that you need to take a good hard look at how and when you consume electricity. Do not just assume that because you can have solar panels on the roof of your house, you are automatically helping the environment.

From an environmental perspective, if you wish to get the very best out of a grid-tie system, you should try to achieve the following:

- Use the power you generate for yourself
- Use solar energy for high load applications such as clothes washing
- Reduce your own power consumption from the grid during times of peak demand

Environmental efficiency: comparing supply and demand

There is an online calculator that will allow you to map your electricity usage over a period of a year and compare it with the amount of sunlight available to your home. Designed specifically for grid-tie installations, this calculator allows you to see how close a fit solar energy is in terms of supply and demand.

Whilst this online calculator is no substitute for a detailed electrical usage survey and research into the exact source of the electricity supplied to you at your location, it will give you a good indication of the likely environmental performance of a solar energy system.

To use this online calculator, you will need to collate information about your electricity usage for each month of the year. You will usually find this information on your electricity bill or by contacting your electricity provider. Then visit *www.SolarElectricityHandbook.com*, follow the links to the Grid-Tie Solar Calculator in the Online Calculators section and fill in the online questionnaire.

In conclusion

- Solar electricity can be a great source of power where your power requirements are modest, there is no other source of electricity easily available and you have a good amount of sunshine
- Solar electricity is not the same as solar heating
- Solar panels absorb photons from sunlight to generate electricity
- Direct sunlight generates the most electricity. Dull days still generate some power
- Solar electricity is unlikely to generate enough electricity to power the average family home, unless major economies in the household power requirements are made first
- Larger solar electric systems have a comparatively high capital cost, but the ongoing maintenance costs are very low
- Smaller solar electric system can actually be extremely cost-effective to buy and install, even when compared to a conventional electricity supply
- It can be much cheaper using solar electricity at a remote building, rather than connecting it to a conventional grid electricity supply
- Stand-alone solar energy systems can have a big environmental benefit if they negate the need for a connection to grid power
- Grid-tie solar energy systems have an environmental benefit in sunny climates where typical electricity usage patterns are similar to the supply of sunlight
- In colder regions, where electricity usage is highest when sunlight is in short supply, the environmental benefits are less certain

A Brief Introduction to Electricity

Before we can start playing with solar power, we need to talk about electricity. To be more precise, we need to talk about voltage, current, resistance, power and energy.

Having these terms clear in your head will help you to understand your solar system. It will also give you confidence that you are doing the right thing when it comes to designing and installing your system.

Don't panic

If you have not looked at electrics since you were learning physics at school, some of the principles of electricity can be a bit daunting to start with. Do not worry if you do not fully grasp everything on your first read through.

There are a few calculations that I show on the next few pages, but I am not expecting you to remember them all! Whenever I use these calculations later on in the book, I show all my workings and, of course, you can refer back to this chapter as you gain more knowledge on solar energy.

Furthermore, the website that accompanies this book includes a number of online tools that you can use to work through most of the calculations involved in designing a solar electric system. You will not be spending hours with a slide-rule and reams of paper working all this out by yourself.

A brief introduction to electricity

When you think of *electricity*, what do you think of? Do you think of a battery that is storing electricity? Do you think of giant overhead pylons transporting electricity? Do you think of power stations that are generating electricity? Or do you think of a device like a kettle or television set or electric motor that is consuming electricity?

The word *electricity* actually covers a number of different physical effects, all of which are related but distinct from each other. These effects are electric charge, electric current, electric potential and electromagnetism:

- An **electric charge** is a build-up of electrical energy. It is measured in coulombs. In nature, you can witness an electric charge in static electricity or in a lightning strike. A battery stores an electric charge
- An **electric current** is the flow of an electric charge, such as the flow of electricity through a cable. It is measured in amps
- An **electric potential** refers to the potential difference in electrical energy between two points, such as between the positive tip and the negative tip of a battery. It is measured in volts. The greater the electric potential (volts), the greater capacity for work the electricity has

- **Electromagnetism** is the relationship between electricity and magnetism, which enables electrical energy to be generated from mechanical energy (such as in a generator) and enables mechanical energy to be generated from electrical energy (such as in an electric motor)

How to measure electricity

Voltage refers to the potential difference between two points. A good example of this is an AA battery: the voltage is the difference between the positive tip and the negative end of the battery. Voltage is measured in *volts* and has the symbol 'V'.

Current is the flow of electrons in a circuit. Current is measured in *amps* (A) and has the symbol 'I'. If you check a power supply, it will typically show the current on the supply itself.

Resistance is the opposition to an electrical current in the material the current is flowing through. Resistance is measured in *ohms* and has the symbol 'R'.

Power measures the rate of energy conversion. It is measured in *watts* (W) and has the symbol 'P'. You will see watts advertised when buying a kettle or vacuum cleaner: the higher the wattage, the more power the device consumes and the faster (hopefully) it does its job.

Energy refers to the capacity for work: power multiplied by time. Energy has the symbol 'E'. Energy is usually measured in *joules* (a joule equals one watt-second), but electrical energy is usually shown as *watt-hours* (Wh), or *kilowatt-hours* (kWh), where 1 kWh = 1,000 Wh.

The relationship between volts, amps, ohms, watts and watt-hours

Volts

$$\text{Current x Resistance} = \text{Volts}$$

$$\text{I x R} = \text{V}$$

Voltage is equal to current multiplied by resistance. This calculation is known as Ohm's Law. As with power calculations, you can express this calculation in different ways. If you know volts and current, you can calculate resistance. If you know volts and resistance, you can calculate current:

$$\text{Volts} \div \text{Resistance} = \text{Current}$$
$$\text{V} \div \text{R} = \text{I}$$

$$\text{Volts} \div \text{Current} = \text{Resistance}$$
$$\text{V} \div \text{I} = \text{R}$$

Power

$$\text{Volts x Current} = \text{Power}$$
$$V \times I = P$$

Power is measured in watts. It equals volts times current. A 12-volt circuit with a 4-amp current equals 48 watts of power (12 x 4 = 48).

Based on this calculation, we can also work out voltage if we know power and current, and current if we know voltage and power:

$$\text{Power} \div \text{Current} = \text{Volts}$$

$$P \div I = V$$

Example: A 48-watt motor with a 4-amp current is running at 12 volts.

$$48 \text{ watts} \div 4 \text{ amps} = 12 \text{ volts}$$

$$\text{Current} = \text{Power} \div \text{Volts}$$

$$I = P \div V$$

Example: a 48-watt motor with a 12-volt supply requires a 4-amp current.

$$48 \text{ watts} \div 12 \text{ volts} = 4 \text{ amps}$$

Power (watts) is also equal to the square of the current multiplied by the resistance:

$$\text{Current}^2 \times \text{Resistance} = \text{Power}$$

$$I^2 \times R = P$$

Energy

Energy is a measurement of power over a period of time. It shows how much power is used, or generated, by a device, typically over a period of an hour. In electrical systems, it is measured in watt-hours (Wh) and kilowatt-hours (kWh).

A device that uses 50 watts of power, has an energy demand of 50Wh per hour. A solar panel that can generate 50 watts of power per hour, has an energy creation potential of 50Wh per hour.

However, because solar energy generation is so variable, based on temperature, weather conditions, the time of day and so on, a new figure is now often shown specifically for solar systems: a watt-peak (Wp) rating.

A watt-peak rating shows how much power can be generated by a solar panel at its peak rating. It has been introduced to highlight the fact that the amount of energy a solar panel can generate is variable and to remind consumers that a solar panel rated at 50 watts is not going to be producing 50 watt-hours of energy every single hour of every single day.

A word for non-electricians

Realistically, if you are new to electrical systems, you should not be planning to install a big solar energy system yourself. If you want a low-voltage system to mount to the roof of a boat, garden shed or barn, or if you want to play with the technology and have some fun, then great: this book will tell you everything you need to know. However, if the limit of your electrical knowledge is wiring a plug or replacing a fuse, you should not be thinking of physically wiring and installing a solar energy system yourself without learning more about electrical systems and electrical safety first.

Furthermore, if you are planning to install a solar energy system to the roof of a house, be aware that in many parts of the world you need to have electrical qualifications in order to carry out even simple household wiring.

That does not mean that you cannot specify a solar energy system, calculate the size you need and buy the necessary hardware for a big project. It does mean that you are going to need to employ a specialist to check your design and carry out the installation.

In conclusion

- Understanding the basic rules of electricity makes it much easier to put together a solar electric system
- As with many things in life, a bit of theory makes a lot more sense when you start applying it in practice
- If this is your first introduction to electricity, you may find it useful to run through it a couple of times
- You may also find it useful to bookmark this section and refer back to it as you read on
- You will also find that, once you have learned a bit more about solar electric systems, some of the terms and calculations will start to make a bit more sense.
- If you are not an electrician, be realistic in what you can achieve. Electrics can be dangerous and you do not want to get it wrong. You can do most of the design work yourself, but you are going to need to get a specialist in to check your design and carry out the installation.

The Four Configurations for Solar Power

There are four different configurations you can choose from when creating a solar electricity installation. These are stand-alone (sometimes referred to as *off-grid*), grid-tie, grid-tie with power backup (also known as *grid interactive*) and grid fallback.

Here is a brief introduction to these different configurations:

Stand-alone/off-grid

Worldwide, stand-alone solar photovoltaic installations are the most popular type of solar installation there is. It is what solar photovoltaics were originally created for: to provide power at a location where there is no other source easily available.

Whether it is powering a shed light, providing power for a pocket calculator or powering a complete off-grid home, stand-alone systems fundamentally all work in the same way: the solar panel generates power, the energy is stored in a battery and then used as required.

In general, stand-alone systems are comparatively small systems, typically with a peak power generation of under one kilowatt.

Almost everyone can benefit from a stand-alone solar system for something, even if it is something as mundane as providing an outside light somewhere. Even if you are planning on something much bigger and grander, it is often a good idea to start with a very small and simple stand-alone system first. Learn the basics and then progress from there.

Examples of simple stand-alone systems

The vending machine
ByBox is a manufacturer of electronic lockers. These are typically used for left luggage at railway stations or at airports, or situated at shopping malls or fuel stations and used as part of a delivery service for people to collect internet deliveries, so they do not need to wait at home.

One of the biggest issues with electronic lockers has often been finding suitable locations to place them where a power source is available. ByBox overcame this issue by building an electronic locker with a solar roof to provide permanent power to the locker.

The solar roof provides power to a set of batteries inside the locker. When not in use, the locker itself is in standby mode, thereby consuming minimal power. When a customer wishes to use the locker, they press the START button and use the locker as normal.

The benefit to ByBox has been twofold: they can install a locker bank in any location, without any dependence on a power supply. Secondly, the cost of the solar panels and controllers is often less than the cost of installing a separate electricity supply, even if there is one nearby.

Recreational vehicles
Holidaying with recreational vehicles or caravans is on the increase, and solar energy is changing the way people are going on holiday.

In the past, most RV owners elected to stay on larger sites, which provided access to electricity and other facilities. As recreational vehicles themselves become more luxurious, however, people are now choosing to travel to more remote locations and live entirely 'off-grid', using solar energy to provide electricity wherever they happen to be. Solar is being used to provide all the comforts of home, whilst offering holidaymakers the freedom to stay wherever they want.

Grid-tie

Grid-tie is gaining popularity in Europe and the United States. This is due to the availability of grants to reduce the installation costs and the ability to earn money by selling electricity back into the electricity companies through a *feed-in tariff*.

Feed-in tariff schemes vary around the world and are not available everywhere. Where they exist, your local electricity company buys electricity from solar producers at an agreed rate per kilowatt-hour. In some countries, this price has been set at an inflated rate by government in order to encourage people to install solar. In other countries and regions, the price is agreed by the electricity companies themselves.

In a grid-tie system, your home runs on solar power during the day. Any surplus energy that you produce is then fed into the grid. In the evenings and at night, when your solar energy system is not producing electricity, you then buy your power from the electricity companies in the usual way.

The benefit of grid-tie solar installations is that they reduce your reliance on the big electricity companies and ensure that more of your electricity is produced in an environmentally efficient way.

One disadvantage of most grid-tie systems is that if there is a power cut, power from your solar array is also cut.

Grid-tie can work especially well in hot, sunny climates, where peak demand for electricity from the grid often coincides with the sun shining, thanks to the high power demand of air conditioning units. Grid-tie also works well where the owners use most of the power themselves.

An example of a grid-tie system

Si Gelatos is a small Florida-based ice-cream manufacturer. In 2007, they installed solar panels on the roof of their factory to provide power and offset some of the energy used in running their cold storage facility.

"Running industrial freezers is extremely expensive and consumes a lot of power," explains Dan Foster of Si Gelatos. "Realistically, we could not hope to generate all of the power from solar, but we felt it was important to reduce our overall power demand and solar allowed us to do that."

Cold storage facilities consume most of their power during the day in the summer, when solar is running at its peak. Since installing solar power, Si Gelatos has seen its overall energy consumption drop by 40% and now hardly takes any power from the utilities during peak operating times.

"Solar has done three things for our business," says Dan. "Firstly, it is a very visible sign for our staff that we are serious about the environment. This in turn has made our employees more aware that they need to do their bit by making sure lights and equipment are switched off when they are not needed. Secondly, it shows our customers that we care for the environment, which has definitely been good for goodwill and sales. Thirdly, and most importantly, we're genuinely making a real contribution to the environment, by reducing our electricity demand at the time of day when everyone else's demand for electricity is high as well."

Grid-tie with power backup (grid interactive)

Grid-tie with power backup – also known as a *grid interactive* system – combines a grid-tie installation with a bank of batteries.

As with grid-tie, the concept is that you use power from your solar array when the sun shines and sell the surplus to the power companies. Unlike a standard grid-tie system, however, a battery bank provides contingency for power cuts so that you can continue to use power from your system.

Typically, you would set up 'protected circuits' within your building that will continue to receive power during a power outage. This ensures that essential power remains available for running lights, refrigeration and heating controllers, for example, whilst backup power is not wasted on inessential items such as televisions and radios.

If there is a potential for main power to be lost for several days, it is also possible to design a system to incorporate other power generators into a grid interactive system, such as a generator. This would allow a grid interactive system to work as a highly efficient *uninterruptable power supply* (UPS) for extended periods of time.

The cost of a grid-tie system with power backup is higher than a standard grid-tie system, because of the additional cost of batteries and battery controllers. Typically, having power backup will add 12–20% of additional costs over a standard grid-tie system.

As with normal grid-tie systems, it is possible to sell surplus power back to the utility companies in some countries, allowing you to earn an income from your solar energy system.

An example of a grid interactive system

Grid interactive systems are gaining popularity with rural farms in the United Kingdom, where even short power blackouts can cause significant disruption.

Traditionally, farms have countered this by using generators to provide light and power. However, between 2009 and 2011, when the UK Government were offering large incentives for installing solar power, many farmers fitted grid interactive systems onto their buildings, providing themselves with an income by selling electricity to the electricity utility companies and giving themselves backup power in case of a power blackout.

The additional cost of installing a grid interactive system over a standard grid-tie system is more than offset by the low running costs and ease of use of the system. Farmers do not need to buy and run generators and the system is almost entirely maintenance-free. This is a big contrast with generator systems, which need to be tested and run regularly in order to ensure they are working effectively.

Grid fallback

Grid fallback is a lesser-known system that makes a lot of sense for smaller household solar power systems. For most household solar installations where solar is being installed for technical or environmental reasons, grid fallback is my preferred solution. Operationally it is effective, it is cost-effective and it is environmentally extremely efficient.

With a grid fallback system, the solar array generates power, which in turn charges a battery bank. Energy is taken from the battery and run through an inverter to power one or more circuits from the distribution panel in the house.

When the batteries run flat, the system automatically switches back to the grid power supply. The solar array then recharges the batteries and the system switches back to solar power.

With a grid fallback system, you do not sell electricity back to the electricity companies. All the power that you generate, you use yourself. This means that some of the grants that are available for solar installations in some countries may not be available to you. It also means that you cannot benefit from selling your electricity back to the electricity companies.

For this reason, grid fallback makes more sense in countries where there is no feed-in tariff available, such as India, or in countries like Australia that have financial incentives available for both grid-tied and off-grid systems.

Grid fallback systems provide most of the benefits of a grid interactive system, with the additional benefit that you use your own power when you need it, rather than when the sun is shining. This reduces your reliance on external electricity supplies during peak load periods, which ensures that your system has an overall environmental benefit.

The other significant benefit of a grid fallback system is cost: you can genuinely build a useful grid fallback system to power one or more circuits within a house for a very small investment and expand it as budget allows. I have seen grid fallback systems installed for under £350 ($560), providing a useful amount of power for a home. In comparison, even a very modest grid-tie system costs several thousands of pounds.

There is a crossover point where a grid-tie system works out more cost-effective than a grid fallback system. At present, that crossover point is around the 1kWh mark: if your system is capable of generating more than 1kW of electricity per hour, a grid-tie system may be more cost-effective. If your system generates less than 1kW of electricity per hour, a grid fallback system is almost certainly cheaper.

Unless you are looking to invest a significant amount of money on a larger grid-tie system in order to produce more than 1 kW of power per hour, or if you want to take advantage of feed-in tariffs, a grid fallback solution is certainly worth investigating as an alternative.

An example of a grid fallback system

Back in 2001, Colin Metcalfe installed a solar panel onto the roof of his garage, in order to charge an old car battery, which in turn powered a single light and a small inverter. After a power cut that winter, Colin decided to expand his system in order to provide basic power to his house.

"I wanted to ensure I always had enough power in my home to power lights and to ensure my heating system would work," explained Colin. "I have gas heating, but the controllers are all electric, which means that if there is a power cut, I have no heating at all. In addition, I liked the idea of free electricity that was generated in an environmentally friendly way."

Colin upgraded his system bit by bit, as funds allowed. "An electrician fitted a new distribution panel (consumer unit) for my essential circuits, and this was connected up to the main panel via an automatic transfer switch. Then I added additional solar panels and batteries over the years as I could afford them."

This automatic transfer switch meant the essential circuits would receive power from the solar array or the batteries while power was available, but switch back to utility power when the batteries ran flat. Originally, the system provided around half the power he needed, but as he has added to the system, more and more of his power now comes from his solar array. "Today I have around 1.4kW of solar panels on the roof of my garage," says Colin. "They look a bit odd as no two panels are alike, as I have bought them bit by bit as funds allow, but they now provide all the power I need around the year for all my essential circuits."

Grid failover

Alternatively, you can configure a grid fallback system as a *grid failover* system.

A grid failover system kicks in when there is a power failure from your main electricity supply. In effect, it is an uninterruptable power supply, generating its power from solar energy.

The benefit of this configuration is that if you have a power cut, you have contingency power. The disadvantage of this configuration is that you are not using solar power for your day-to-day use.

Although rare in Europe and America, grid failover systems used to be more common in countries where power failures are commonplace. In Africa and in many parts of Asia, grid failover systems reduce the reliance on power generators for lighting and basic electricity needs.

However, in most cases, customers have found that a grid fallback or grid interactive system is more suitable for their needs. I am aware of two grid failover systems that have been installed in the past: both of these have since been reconfigured as grid fallback systems.

How grid-tie systems differ from stand-alone

Generally, stand-alone and smaller grid fallback systems run at low voltages, typically between 12 and 48 volts. This is because batteries are low-voltage units and so building a stand-alone system at a low voltage is a simple, flexible and safe approach.

Grid-tie systems tend to be larger installations, often generating several kilowatts of electricity each hour. As the electricity is required as a high-voltage supply, it is more efficient to connect multiple solar panels together to produce a high voltage circuit, rather than use an inverter to step up the voltage. This high-voltage DC power is then converted into an AC current by a suitable grid-tie inverter.

Grid-tie systems either link multiple solar panels together to produce a solar array voltage of several hundred volts before running to the inverter, or have a small inverter connected to each solar panel to create a high-voltage AC supply from each panel.

The benefit of this high voltage is efficiency. There is less power loss running high-voltage, low-current electricity through cables from the solar array.

For stand-alone battery-based systems, low-voltage is the best solution, as the battery banks tend to work better as low-voltage energy stores. For grid-tie systems where the energy is not being stored in a battery bank, the higher-voltage systems are the best solution. Neither approach is inherently 'better': it all depends on the type of system you are designing.

In conclusion

- Solar can be used in a number of different ways and for many different applications
- Stand-alone systems are the simplest and easiest to understand. They tend to be comparatively small systems, providing power where no other power source is easily available
- With grid-tie, your solar energy system generates electricity that is then used normally. Any excess electricity production is exported onto the grid
- Grid-tie with power backup (also known as grid interactive) provides you with the benefits of a grid-tie system with the added benefit that power remains available even if electricity to your area is cut off
- Grid fallback systems have more in common with stand-alone systems than grid-tie systems. In design they are very similar to stand-alone systems, with an inverter running from a bank of batteries and an automatic transfer switch to switch power between the solar energy system and the grid power supply
- Grid failover systems are comparatively rare now, but provide uninterruptable power supplies using solar as the backup source
- Grid-tie systems have a different design to stand-alone systems. They tend to be high-voltage systems, whereas stand-alone systems run at much lower voltages

Components of a Solar Electric System

Before I get into the detail about planning and designing solar electric systems, it is worth describing all the different components of a system and explaining how they fit together. Once you have read this chapter, you will have a reasonable grasp of how a solar energy system fits together.

I deliberately do not go into much detail at this stage: all I am doing is providing an overview for now. The detail can come later.

Solar panels

The heart of a solar electric system is the solar panel itself. There are various types of solar panel and I will describe them all in detail later on.

Solar panels or, more accurately, *photovoltaic* solar panels, generate electricity from the sun. The more powerful the sun's energy, the more power you get, although solar panels continue to generate small amounts of electricity in the shade.

Most solar panels are made up of individual solar cells, connected together. A typical solar cell will only produce around half a volt, so by connecting them together in series inside the panel, a more useful voltage is achieved.

Most solar panels are rated as 12-volt solar panels, although higher-voltage panels are also available. A 12-volt solar panel produces around 14–18 volts when put under load. This allows a single solar panel to charge up a 12-volt battery.

Incidentally, if you connect a voltmeter up to a solar panel when it is not under load, you may well see voltage readings of up to 26 volts. This is normal in an 'open circuit' on a solar panel. As soon as you connect the solar panel into a circuit, this voltage level will drop to around 14–18 volts.

Solar panels can be linked together to create a *solar array*. Connecting multiple panels together allows you to produce a higher current or to run at a higher voltage:

- Connecting the panels *in series* allows a solar array to run at a higher voltage. Typically, 24 volts or 48 volts in a stand-alone system, or up to several hundred volts in a grid-tie system
- Connecting the panels *in parallel* allows a solar array to produce more power while maintaining the same voltage as the individual panels
- When you connect multiple panels together, the power of the overall system increases, irrespective of whether they are connected in series or in parallel

In a solar array where the solar panels are connected in series (as shown in the diagram over the page), you add the voltages of each panel together and add the wattage of each panel together to calculate the maximum amount of power and voltage the solar array will generate.

A solar array made of four solar panels connected in series. If each individual panel is rated as a 12-volt, 12-watt panel, this solar array would be rated as a 48-volt, 48-watt array with a 1 amp current.

In a solar array where the panels are connected in parallel (as shown in the diagram below), you take the *average* voltage of all the solar panels and you add the wattage of each panel to calculate the maximum amount of power the solar array will generate.

A solar array made of four solar panels connected in parallel. With each panel rated as a 12-volt, 12-watt panel, this solar array would be rated as a 12-volt, 48-watt array with a 4 amp current.

I will go into more detail later about choosing the correct voltage for your system.

Batteries

Except in a grid-tie system, where the solar array connects directly to an inverter, solar panels rarely power electrical equipment directly. This is because the amount of power the solar panel collects varies depending on the strength of sunlight. This makes the power source too variable for most electrical equipment to cope with.

In a grid-tie system, the inverter handles this variability: if demand outstrips supply, you will get power from both the grid and your solar system. For a stand-alone or a grid fallback system, batteries store the energy and provide a constant power source for your electrical equipment.

Typically, this energy is stored in 'deep cycle' lead acid batteries. These look similar to car batteries but have a different internal design. This design allows them to be heavily discharged and recharged several hundred times over.

Most lead acid batteries are 6-volt or 12-volt batteries and, like solar panels, these can be connected together to form a larger *battery bank*. Like solar panels, multiple batteries used in series increase the capacity and the voltage of a battery bank. Multiple batteries connected in parallel increase the capacity whilst keeping the voltage the same.

Controller

If you are using batteries, your solar electric system is going to require a controller in order to manage the flow of electricity (the current) into and out of the battery.

If your system overcharges the batteries, this will damage and eventually destroy them. Likewise, if your system completely discharges the batteries, this will quite rapidly destroy them. A solar controller prevents this from happening.

There are a few instances where a small solar electric system does not require a controller. An example of this is a small 'battery top-up' solar panel that is used to keep a car battery in peak condition when the car is not being used. These solar panels are too small to damage the battery when the battery is fully charged.

In the majority of instances, however, a solar electric system will require a controller in order to manage the charge and discharge of batteries and keep them in good condition.

Inverter

The electricity generated by a solar electric system is direct current (DC). Electricity from the grid is high-voltage alternating current (AC).

If you are planning to run equipment that runs from grid-voltage electricity from your solar electric system, you will need an inverter to convert the current from DC to AC and convert the voltage to the same voltage as you get from the grid.

Traditionally, there is usually one central inverter in a solar system, either connecting directly to the solar array in a grid-tie system, or to the battery pack in an off-grid system. A more recent invention has been the micro inverter. Micro-inverters are connected to individual solar panels so that each individual panel provides a high-voltage alternating current.

Solar panels with micro-inverters are typically only used with grid-tie systems and are not suitable for systems with battery backup. For grid-tie systems, they do offer some significant benefits over the more traditional 'big box' inverter, although the up-front cost is currently higher.

Inverters are a big subject all on their own. I will come back to describe them in much more detail later on in the book.

Electrical devices

The final element of your solar electric system is the devices you plan to power. Theoretically, anything that you can power with electricity can be powered by solar. However, many electrical devices are very power hungry, which makes running them on solar energy very expensive!

Of course, this may not be so much of an issue if you are installing a grid-tie system: if you have very energy-intensive appliances that you only use for short periods, the impact to your system is low. In comparison, running high-power appliances on an off-grid system means you have to have a more powerful off-grid solar energy system to cope with the peak demand.

Low-voltage devices

Most off-grid solar systems run at low voltages. Unless you are planning a pure grid-tie installation, you may wish to consider running at least some of your devices directly from your DC supply rather than running everything through an inverter. This has the benefit of greater efficiency.

Thanks to the caravanning and boating communities, lots of equipment is available to run from a 12-volt or 24-volt supply: light bulbs, refrigerators, ovens, kettles, toasters, coffee machines, hairdryers, vacuum cleaners, televisions, radios, air conditioning units, washing machines and laptop computers are all available to run on 12-volt or 24-volt power.

In addition, thanks to the recent uptake in solar installations, some specialist manufacturers are building ultra low-energy appliances, such as refrigerators, freezers and washing machines, specifically for people installing solar and wind turbine systems.

You can also charge up most portable items such as MP3 players and mobile phones from a 12-volt supply.

High-voltage devices

If running everything at low voltage is not an option, or if you are using a grid-tie system, you use an inverter to run your electrical devices.

Connecting everything together

A stand-alone system

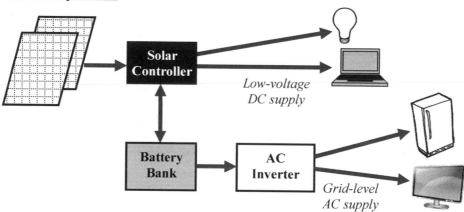

The simplified block diagram above shows a simple stand-alone solar electric system. Whilst the detail will vary, this design forms the basis of most stand-alone systems and is typical of the installations you will find in caravans, boats and buildings that do not have a conventional power supply.

This design provides both low-voltage DC power for running smaller electrical devices and appliances such as laptop computers and lighting, plus a higher-voltage AC supply for running larger devices such as larger televisions and kitchen appliances.

In this diagram, the arrows show the flow of current. The solar panels provide the energy, which is fed into the solar controller. The solar controller charges the batteries. The controller also supplies power to the low-voltage devices, using either the solar panels or the batteries as the source of this power.

The AC inverter takes its power directly from the battery and provides the high-voltage AC power supply.

A grid-tie system using a single central inverter

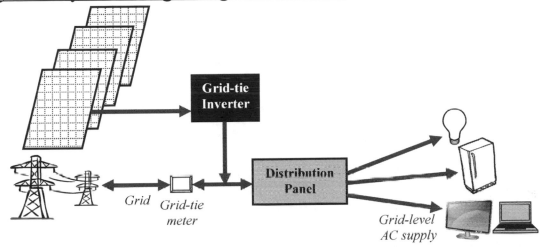

This simplified block diagram shows a simple grid-tie system, typical of the type installed in many homes today. The solar panels are connected to the grid-tie inverter, which feeds the energy into the main supply. Electricity can be used by devices in the building or fed back out onto the grid, depending on demand.

The grid-tie inverter monitors the power feed from the grid. If it detects a power cut, it also cuts power from the solar panels to ensure that no energy is fed back out onto the grid.

The grid-tie meter monitors how much energy is taken from the grid and how much is fed back into the grid using the solar energy system.

A grid-tie system using multiple micro-inverters

A grid-tie system using micro-inverters is similar to the one above, except that each solar panel is connected to its own inverter, and the inverters themselves are daisy-chained together, converting the low-voltage DC power from each solar panel into a high-voltage AC power supply.

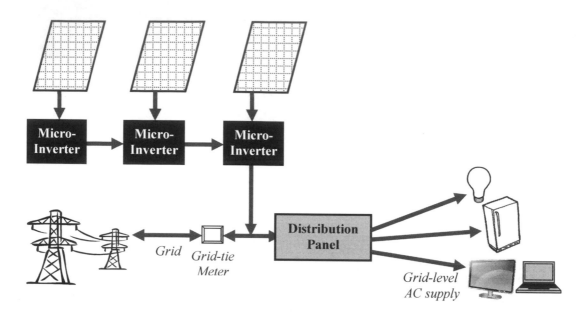

In conclusion

- There are various components that make up a solar electric system
- Multiple solar panels can be joined together to create a more powerful *solar array*.
- In a stand-alone system, the electricity is stored in batteries to provide an energy store and provide a more constant power source. A controller manages the batteries, ensuring the batteries do not get overcharged by the solar array and are not over-discharged by the devices taking current from them
- An inverter takes the DC current from the solar energy system and converts it into a high-voltage AC current that is suitable for running devices that require grid power
- Generally, it is more efficient to use the electricity as a DC supply than an AC supply

The Design Process

No matter what your solar energy system is for, there are seven steps in the design of every successful solar electric installation:

- Scope the project
- Calculate the amount of energy you need
- Calculate the amount of solar energy available
- Survey your site
- Size up the solar electric system
- Select the right components and work out full costs
- Produce the detailed design

The design process can be made more complicated, or simplified, based on the size of the project. If you are simply installing an off-the-shelf shed light, for instance, you can probably complete the whole design in around twenty minutes. If, on the other hand, you are looking to install a solar electric system in a business to provide emergency site power in the case of a power cut, your design work is likely to take considerably more time.

Whether your solar electric system is going to be large or small, whether you are buying an off-the-shelf solar lighting kit or designing something from scratch, it is worth following this basic design process every time. This is true even if you are installing an off-the-shelf system. This ensures that you will always get the best from your system and will provide you with the reassurance that your solar energy system will achieve everything you need it to do.

Short-cutting the design work

Having said that doing the design work is important, there are some useful online tools to help make the process as easy as possible.

Once you have scoped your project, the Solar Electricity Handbook website (*www.SolarElectricityHandbook.com*) includes a number of online tools and calculators that will help you carry out much of the design work.

The solar irradiance tables and solar angle calculators will allow you to work out how much solar energy is available at your location, whilst the off-grid project analysis and grid-tie project analysis questionnaires will each generate and e-mail to you a full report for your proposed system, including calculating the size of system you require and providing a cost estimate.

Of course, there is a limit to how much a set of online solar tools can help you in isolation, so you will still need to carry out a site survey and go through components selection and detailed design yourself, but these tools will allow you to try several different configurations and play out 'what if' scenarios quickly and easily.

Incidentally, whilst some of these tools ask you for an e-mail address (in order to send you your report), your e-mail address is not stored anywhere on the system. Other than the report that you request, you will never receive unsolicited e-mails because of entering your e-mail address.

Solar energy and emotions

Design can often seem to be a purely analytical and rational process. It should not be. All great designs start with a dream.

For many people, choosing solar energy is often an emotional decision: they want a solar energy system for reasons other than just the purely practical. Some people want solar energy because they want to 'do their bit' for the environment, others want the very latest technology, or want to use solar simply because it can be done. Others want solar energy because they see the opportunity to earn money. I suspect that for most homeowners, the reasons are a combination of the above.

It is so important that the emotional reasons for wanting something are not ignored. We are not robots. Our emotions should be celebrated, not suppressed: the Wright brothers built the first aircraft because they wanted to reach the sky. NASA sent a man to the moon because they wanted to go further than anyone had ever done before. Neither undertaking could be argued as purely rational; they were the results of big dreams.

It is important to acknowledge that there are often hidden reasons for wanting solar energy. Sadly, these reasons often do not make it down onto a sheet of paper in a design document or onto a computer spreadsheet. Sometimes, the person making the decision for buying solar energy is secretly worried that if they voice their dreams, they will appear in some way irrational.

The reality is that it is often a good thing if there is an emotional element to wanting a solar energy system. By documenting these reasons, you will end up with a better solution. For instance, if the environmental benefits are top of your agenda, you will use your solar energy system in a different way to somebody who is looking at solar purely as a business investment.

By acknowledging these reasons and incorporating them into the design of your system, you will end up with a far better system. Not only will you have a system that works in a practical sense, it will also achieve your dream.

In conclusion

- No matter how big or small your project, it is important to design it properly
- There are online tools available to help you with the calculations and to speed up the work
- Do not ignore the emotional reasons for wanting a solar energy system. You are a human being: you are allowed to dream

Scoping the Project

As with any project, before you start, you need to know what you want to achieve. In fact, it is one of the most important parts of the whole project. Get it wrong and you will end up with a system that will not do what you need it to.

It is usually best to keep your scope simple to start with. You can then flesh it out with more detail later.

Here are some examples of a suitable scope:

- To power a light and a burglar alarm in a shed on an allotment
- To provide power for lighting, a kettle, a radio and some handheld power tools in a workshop that has no conventional electrical connection
- To provide enough power for lighting, refrigeration and a TV in a holiday caravan
- To provide lighting and power to run four laptop computers and the telephone system in an office during a power cut
- To charge up an electric bike between uses
- To provide an off-grid holiday home with its entire electricity requirements
- To reduce my carbon footprint by producing electricity for all my home requirements
- To run an electric car entirely on solar energy

From your scope, you can start fleshing this out to provide some initial estimates on power requirements.

As mentioned in the previous chapter, I have created two online Solar Project Analysis tools, one for grid-tie systems and one for off-grid systems. You can find both of these tools on my website www.SolarElectricityHandbook.com. You will still need to collect the basic information to work with, but all the hard work is done for you. This tool will produce a complete project scope, work out the likely performance of your solar energy system and provide some ballpark cost estimates.

For the purpose of these next few chapters, I am going to use the example of providing a small off-grid holiday home with its entire electricity requirements.

This is a big project. In reality if you have little or no experience of solar electric systems or household electrics you would be best starting with something smaller. Going completely off-grid is an ambitious project, but for the purposes of teaching solar electric system design, it is a perfect example: it requires a detailed design that covers all of the aspects of designing a solar energy system.

Designing grid-tie or grid fallback systems

For our sample project, grid-tie is not an option as we are using solar power as an alternative to connecting our site to the electricity grid.

However, grid-tie is becoming a popular option, especially in the southern states of the United States and in European countries like Spain, Germany and the United Kingdom where generous government subsidies and feed-in tariffs are available.

In terms of scoping the project, it makes little difference whether you are planning a grid-tie or grid fallback system or not: the steps you need to go through are the same. The only exception, of course, is that you do not need to take into account battery efficiencies with grid-tie.

The biggest difference with a grid-tie or grid fallback system is that you do not have to rely on your solar system providing you with all your electricity requirements: you will not be plunged into darkness if you use more electricity than you generate.

This means that you can start with a small grid-tie or grid fallback system and expand it later on as funds allow.

Despite that, it is still a good idea to go through a power analysis as part of the design. Even if you do not intend to produce all the power you need with solar, having a power analysis will allow you to benchmark your system and will help you size your grid-tie system if you aim to reduce your carbon footprint by providing the electricity companies with as much power as you buy back.

Most grid-tie systems are sized to provide more power than you need during the summer and less than you need during the winter. Over a period of a year, the aim is to generate as much power as you use, although on a month-by-month basis this may not always be the case.

Many solar companies claim that this then provides you with a 'carbon neutral' system: you are selling your excess power to the electricity companies and then buying the same amount of electricity back when you need it.

If this is what you are planning to do with your grid-tie system, your scope is much simpler. You need to get your electricity bills for the past year and make a note of how much electricity you have used over the year. Then divide this figure by the number of days in the year to work out a daily energy usage and ensure your system generates this as an average over the period of a year.

Because you are not generating enough electricity during the winter months in a carbon neutral grid-tie system, you need fewer solar panels than you would need to create an entirely stand-alone system.

Comparing supply with demand

If you are designing a grid-tie system, it can be interesting to compare the supply of solar energy with your electricity usage pattern. By comparing supply with demand, you can see how closely solar energy production matches your own usage and this, in turn, can be used as an indicator to identify how environmentally beneficial solar energy is for you.

To do this, you will need to ascertain your monthly electricity usage for each month of the year. Your electricity supplier may already provide you with this information on your electricity bill. If not, you should be able to request this from them.

Once you have this information, visit *www.SolarElectricityHandbook.com* and fill in the Grid-Tie Solar Project Analysis, including your individual monthly consumption figures. In the report that is e-mailed to you, you will see a chart that allows you to see how closely your electricity usage maps onto solar energy production.

This report will also provide you with an approximate estimate for the carbon footprint for each kilowatt-hour of electricity you produce from your solar array, based on the production and installation of your solar array and the likely amount of energy that it will generate during its lifetime.

Based on this, it is possible to see whether installing solar energy is likely to produce *real-world* environmental savings.

Fleshing out the scope

Now we know the outline scope of our project, we need to quantify exactly what we need to achieve and work out some estimates for energy consumption.

Our holiday home is a small two-bedroom cottage with a solid fuel cooker and boiler. The cost of connecting the cottage to the grid is £4,100 (around $6,520) and I suspect that solar electric power could work out significantly cheaper.

The cottage is mainly used in the spring, summer and autumn, with only a few weekend visits during the winter.

Electricity is required for lighting in each room, plus a courtesy light in the porch, a fridge in the kitchen and a small television in the sitting room. There also needs to be surplus electricity for charging up a mobile phone or MP3 player and for the occasional use of a laptop computer.

Now we have decided what devices we need to power, we need to find out how much energy each device needs, and estimate the daily usage of each item.

In order to keep efficiency high and costs low, we are going to work with low-voltage electrics wherever possible. The benefits of using low-volt devices rather than higher grid-voltage are twofold:

- We are not losing efficiency by converting low-volt DC electrics to grid-voltage AC electrics through an inverter.
- Many electronic devices that plug into a grid-voltage socket require a transformer to reduce the power back down to a low DC current, thereby creating a second level of inefficiency

Many household devices, like smaller televisions, music systems, computer games consoles and laptop computers, have external transformers. It is possible to buy transformers that run on 12-volt or 24-volt electrics rather than the AC voltages we get from grid power, and using these is the most efficient way of providing low-voltage power to these devices.

There can be disadvantages of low-voltage configurations, however, and they are not the right approach for every project:

- If running everything at 12–24 volts requires a significant amount of additional rewiring, the cost of carrying out the rewiring can be much higher than the cost of an inverter and a slightly larger solar array

- If the cable running between your batteries and your devices is too long, you will get greater power losses through the cable at lower voltages than you will at higher voltages

If you already have wiring in place to work at grid-level voltages, it is often more appropriate to run a system at grid voltage using an inverter, rather than running the whole system at low voltage. If you have no wiring in place, running the system at 12 or 24 volts is often more suitable.

Producing a power analysis

The next step is to investigate your power requirements by carrying out a power analysis, where you measure your power consumption in watt-hours.

You can find out the wattage of household appliances in one of four ways:

- Check the rear of the appliance, or on the power supply
- Check the product manual
- Measure the watts using a watt meter
- Find a ballpark figure for similar items

Often a power supply will show an output current in amps rather than the number of watts the device consumes. If the power supply also shows the output voltage, you can work out the wattage by multiplying the voltage by the current (amps):

$$Power\ (watts) = Volts\ x\ Current\ (amps)$$
$$P = V\ x\ I$$

For example, if you have a mobile phone charger that uses 1.2 amps at 5 volts, you can multiply 1.2 amps by 5 volts to work out the number of watts: in this example, it equals 6 watts of power. If I plugged this charger in for one hour, I would use 6 watt-hours of energy.

A watt meter is a useful tool for measuring the energy requirements of any device that runs on high-voltage AC power from the grid. The watt meter plugs into the wall socket and the appliance plugs into the watt meter. An LCD display on the watt meter then displays the amount of power the device is using. This is the most accurate way of measuring your true power consumption.

Finding a ballpark figure for similar devices is the least accurate way of finding out the power requirement and should only be done as a last resort. A list of power ratings for some common household appliances is included in Appendix C.

Once you have a list of the power requirements for each electrical device, draw up a table listing each device, whether the device uses 12-volt or grid voltage, and the power requirement in watts.

Then put an estimate in hours for how long you will use each device each day and multiply the watts by hours to create a total watt-hour energy requirement for each item.

You should also factor in any 'phantom loads' on the system. A phantom load is the name given to devices that use power even when they are switched off. Televisions in standby mode are one such example, but any device that has a power supply built into the plug also has a phantom load. These items should be unplugged or switched off at the switch when not in use. However, you may wish to factor in a small amount of power for items in standby mode, to take into account the times you forget to switch something off.

If you have a gas-powered central heating system, remember that most central heating systems have an electric pump and the central heating controller will require electricity as well. A typical central heating pump uses around 60 watts of power a day, whilst a central heating controller can use between 2 and 24 watts a day.

Once complete, your power analysis will look like this:

Device	Voltage	Power (watts)	Hours of use per day	Watt-hours energy
Living room lighting	12V	11W	5	55Wh
Kitchen lighting	12V	11W	2	22Wh
Hallway lighting	12V	8W	½	4Wh
Bathroom lighting	12V	11W	1½	17Wh
Bedroom 1 lighting	12V	11W	1	11Wh
Bedroom 2 lighting	12V	11W	1½	17Wh
Porch light	12V	8W	½	4Wh
Small fridge	12V	12W	24	288Wh
TV	12V	40W	4	160Wh
Laptop computer	12V	40W	1	40Wh
Charging cell phones and MP3 players	12V	5W	4	20Wh
Phantom loads	12V	1W	24	24Wh
Total Energy Requirement a day (watt-hours)				**662Wh**

A word of warning

In the headlong enthusiasm for implementing a solar electric system, it is very easy to underestimate the amount of electricity you need at this stage.

To be sure that you do not leave something out which you regret later, I suggest you have a break at this point. Then return and review your power analysis.

It can help to show this list to somebody else in order to get their input as well. It is very easy to get emotionally involved in your solar project, and having a second pair of eyes can make a world of difference later on.

When you are ready to proceed

We now know exactly how much energy we need to store in order to provide one day of power. For our holiday home example, that equates to 662 watt-hours per day.

There is one more thing to take into account: the efficiency of the overall system.

Batteries, inverters and resistance in the circuits all reduce the efficiency of our solar electric system. We must consider these inefficiencies and add them to our power analysis.

Calculating inefficiencies

Batteries do not return 100% of the energy used to charge them. The *Charge Cycle Efficiency* of the battery measures the percentage of energy available from the battery compared to the amount of energy used to charge it.

Charge cycle efficiency is not a fixed figure, as the efficiency can vary depending on how quickly you charge and discharge the battery. However, most solar applications do not overstress batteries and so the standard charge cycle efficiency figures are usually sufficient.

Approximate charge cycle efficiency figures are normally available from the battery manufacturers. However, for industrial quality 'traction' batteries, you can assume 95% efficiency, whilst gel batteries and leisure batteries are usually in the region of 90%.

If you are using an inverter in your system, you need to factor in the inefficiencies of the inverter. Again, the actual figures should be available from the manufacturer but typically, you will find that an inverter is around 90% efficient.

Adding the inefficiencies to our power analysis

In our holiday home example, there is no inverter. If there were, we would need to add 10% for inverter inefficiencies for every grid-powered device.

We are using batteries. We need to add 5% to the total energy requirement to take charge cycle efficiency into account.

5% of 662 equals 33 watts. Add this to our power analysis, and our total watt-hour requirement becomes 695 watt-hours per day.

When do you need to use the solar system?

It is important to work out at what times of year you will be using your solar electric system most. For instance, if you are planning to use your system full time during the depths of winter, your solar electric system needs to be able to provide all your electricity even during the dull days of winter.

A holiday home is often in regular use during the spring, summer and autumn, but left empty for periods of time during the winter.

This means that, during winter, we do not need our solar electric system to provide enough electricity for full occupancy. We need enough capacity in the batteries to provide enough electricity for, say, the occasional long weekend. The solar array can then recharge the batteries again, once the home is empty.

We might also decide that, if we needed additional electricity in winter, we could have a small standby generator on hand to give the batteries a boost charge.

For the purposes of our holiday home, our system must provide enough electricity for full occupancy from March to October and occasional weekend use from November until February.

Keeping it simple

You have seen what needs to be taken into account when creating a power analysis and calculating the inefficiencies. If you are planning to use the online tools to help you, now is the time to use them.

Visit *www.SolarElectricityHandbook.com* and follow the links to either the Off-Grid or Grid-Tied Solar Project Analysis tools, which can be found in the *Online Calculators* section. This will allow you to enter your devices on the power analysis, select the months you want your system to work and select your location from a worldwide list. The system will automatically e-mail you a detailed solar analysis report with all the calculations worked out for you.

Improving the scope

Based on the work done, it is time to put more detail on our original scope. Originally, our scope looked like this:

> Provide an off-grid holiday home with its entire electricity
> requirements.

Now the improved scope has become:

> Provide an off-grid holiday home with its entire electricity requirements, providing power for lighting, refrigeration, TV, laptop computer and various sundries, which equals 695 watt-hours of electricity consumption per day.

> The system must provide enough power for occupation from March until October, plus occasional weekend use during the winter.

There is now a focus for the project. We know what we need to achieve for a solar electric system to work. Now we need to go to the site and see if what we want to do is achievable.

In conclusion

- Getting the scope right is important for the whole project
- Start by keeping it simple and then flesh it out by calculating the energy requirements for all the devices you need to power
- If you are designing a grid-tie system, you do not need to go into so much detail: you can get the usage information from your electricity company. It is probably included on your electricity bill
- If you are designing a grid-tie system, you can make a reasonable estimate of its environmental benefit by comparing solar energy supply with your demand on a month-by-month basis
- Do not forget to factor in 'phantom loads'
- Because solar electric systems run at low voltages, running your devices at low voltage is more efficient than inverting the voltage to grid levels first
- Thanks to the popularity of caravans and boats, there is a large selection of 12-volt appliances available. If you are planning a stand-alone or grid fallback system, you may wish to use these in your solar electric system rather than less efficient grid-voltage appliances
- Even if you are planning a grid-tie system, it is still useful to carry out a detailed power analysis
- Do not forget to factor in inefficiencies for batteries and inverters
- Take into account the times of year that you need to use your solar electric system
- Once you have completed this stage, you will know what the project needs to achieve in order to be successful

Calculating Solar Energy

The next two chapters are just as useful for people wishing to install a solar hot water system as they are for people wishing to install solar electricity.

Whenever I refer to *solar panel* or *solar array* (multiple solar electric panels) in these two chapters, the information is equally valid for solar electricity and solar hot water.

What is solar energy?

Solar energy is a combination of the hours of sunlight you get at your site and the strength of that sunlight. This varies depending on the time of year and where you live.

This combination of hours and strength of sunlight is called *solar insolation* or *solar irradiance*, and the results can be expressed as watts per square metre (W/m^2) or, more usefully, in kilowatt-hours per square metre spread over the period of a day ($kWh/m^2/day$). One square metre is equal to 9.9 square feet.

Why is this useful?

Photovoltaic solar panels quote the expected number of watts of power they can generate, based on a solar irradiance of 1,000 watts per square metre. This figure is often shown as a watts-peak (Wp) figure and shows how much power the solar panel can produce in ideal conditions. A solar irradiance of 1,000 watts per square metre is what you could expect to receive at solar noon in the middle of summer at the equator. It is not an average reading that you could expect to achieve on a daily basis.

However, once you know the solar irradiance for your area, quoted as a daily average (i.e. the number of kilowatt-hours per square metre per day), you can multiply this figure by the wattage of the solar panel to give you an idea of the daily amount of energy you can expect your solar panels to provide.

Calculating solar irradiance

Solar irradiance varies significantly from one place to another and changes throughout the year. In order to come up with some reasonable estimates, you need irradiance figures for each month of the year for your specific location.

Thanks to NASA, calculating your own solar irradiance is simple. NASA's network of weather satellites has been monitoring the solar irradiance across the surface of the earth for many decades. Their figures have taken into account the upper atmospheric conditions, average cloud cover and surface temperature, and are based on sample readings every three hours for the past quarter of a decade. They cover the entire globe.

For reference, I have compiled this information for different regions across the United States, Canada, Australia, New Zealand, the United Kingdom and Ireland in Appendix B.

The website goes further. We have incorporated solar irradiance charts for every major town and city in every country in the world: simply select your location from a pull-down list of countries and cities and you can view the irradiance figures for your exact area.

Using the information in Appendix B on page 153, here are the solar irradiance figures for London in the United Kingdom, shown on a month-by-month basis. They show the average daily irradiance, based on mounting the solar array flat on the ground:

	Jan	Feb	Mar	Apr	May	Jun	Jul	Aug	Sep	Oct	Nov	Dec
Flat	0.75	1.37	2.31	3.57	4.59	4.86	4.82	4.20	2.81	1.69	0.92	0.60

These figures show how many hours of equivalent midday sun we get over the period of an average day of each month. In the chart above, you can see that in December we get the equivalent of 0.6 of an hour of midday sun (36 minutes), whilst in June we get the equivalent of 4.86 hours of midday sunlight (4 hours and 50 minutes).

Capturing more of the sun's energy

The tilt of a solar panel has an impact on how much sunlight you capture: mount the solar panel flat against a wall or flat on the ground and you will capture less sunlight throughout the day than if you tilt the solar panels to face the sun.

The figures above show the solar irradiance in London, based on the amount of sunlight shining on a single square metre of the ground. If you mount your solar panel at an angle, tilted towards the sun, you can capture more sunlight and therefore generate more power. This is especially true in the winter months, when the sun is low in the sky.

The reason for this is simple: when the sun is high in the sky the intensity of sunlight is high. When the sun is low in the sky the sunlight is spread over a greater surface area:

This diagram shows the different intensity of light depending on the angle of sun in the sky. When the sun is directly overhead, a 1m-wide shaft of sunlight will cover a 1m-wide area on the ground. When the sun is low in the sky – in this example, I am using an angle of 30° towards the sun – a 1m-wide shaft of sunlight will cover a 2m-wide area on the ground. This means the intensity of the sunlight is half as much when the sun is at an angle of 30° compared to the intensity of the sunlight when the sun is directly overhead.

The impact of tilting solar panels on solar irradiance

If we tilt our solar panels towards the sun, it means we can capture more of the sun's energy to convert into electricity. Often the angle of this tilt is determined for you by the angle of an existing roof. However, for every location there are optimal angles at which to mount your solar array, in order to capture as much solar energy as possible.

Using London as an example again, this chart shows the difference in performance of solar panels, based on the angle at which they are mounted. The angles I have shown are flat on the ground, upright against a wall, and mounted at different angles designed to get the optimal amount of solar irradiance at different times of the year (I explain the relevance of these specific angles in a moment):

	Jan	Feb	Mar	Apr	May	Jun	Jul	Aug	Sep	Oct	Nov	Dec
Flat	0.75	1.37	2.31	3.57	4.59	4.86	4.82	4.20	2.81	1.69	0.92	0.60
Upright	1.20	1.80	2.18	2.58	2.70	2.62	2.71	2.80	2.47	2.07	1.43	1.01
38° angle Best year-round tilt	1.27	2.04	2.76	3.67	4.17	4.20	4.25	4.16	3.26	2.41	1.53	1.05
23° angle Best winter tilt	1.30	2.03	2.62	3.34	3.66	3.69	3.76	3.73	3.06	2.37	1.56	1.08
53° angle Best summer tilt	1.19	1.95	2.77	3.84	4.52	4.63	4.66	4.41	3.31	2.33	1.43	0.97
Tilt adjusted each month	1.30	2.05	2.78	3.86	4.70	4.91	4.90	4.46	3.31	2.41	1.56	1.08
	22° tilt	30° tilt	38° tilt	46° tilt	54° tilt	62° tilt	54° tilt	46° tilt	38° tilt	30° tilt	22° tilt	14° tilt

Note: All angles are given in degrees from vertical and are location specific.

Look at the difference in the performance based on the tilt of the solar panel. In particular, look at the difference in performance in the depths of winter and in the height of summer.

It is easy to see that some angles provide better performance in winter; others provide better performance in summer, whilst others provide a good compromise all-year-round solution.

Calculating the optimum tilt for solar panels

Because of the 23½° tilt of the earth relative to the sun, the optimum tilt of your solar panels will vary throughout the year, depending on the season. In some installations, it is feasible to adjust the tilt of the solar panels each month, whilst in others it is necessary to have the array fixed in position.

To calculate the optimum tilt of your solar panels, you can use the following sum:

90° – your latitude = optimum fixed year-round setting

This angle is the optimum tilt for fixed solar panels for all-year-round power generation. This does *not* mean that you will get the maximum power output every single month: it means that across the whole year, this tilt will give you the best compromise, generating electricity all the year round.

Getting the best from solar panels at different times of the year

Depending on when you want to use your solar energy, you may choose to use a different tilt in order to improve power output at a given point in the year. Each month of the year, the angle of the sun in the sky changes by 7.8° – higher in the summer and lower in the winter. By adjusting the tilt of your solar panel to track the sun, you can tweak the performance of your system according to your requirements.

You may want to do this for a number of reasons:

- For a stand alone, off-grid system, you often need to get as much power generation during the winter months as possible to counter the reduction in natural light
- When installing a grid-tie system in a cool climate, where the focus is on reducing your carbon footprint, you can choose to boost your electricity production in winter, to offset the amount of electricity you need to buy when demand is at its highest
- When installing a grid-tie system where the focus is on making the most profit by selling your power, you can choose to tilt your solar panels at a summer setting and thereby produce the maximum amount of energy possible over the course of the year

You can see this monthly optimum angle (rounded to the nearest whole degree) on the bottom row of the table on the previous page.

Optimum winter settings

Here is an example of how you could tweak your system. Performance of a solar system is at its worst during the winter months. However, by tilting your panels to capture as much of the sunlight as possible during the winter, you can significantly boost the amount of power you generate at this time.

Based on the Northern Hemisphere, an optimum winter tilt for solar panels is the optimum angle for November and January. For the Southern Hemisphere, the optimum winter tilt is May and July:

90° – your latitude – 15.6° = optimum winter setting

As you can see from the table on the previous page, if you tilt your solar panels at this angle, you will sacrifice some of your power generation capability during the summer months. However, as you are generating so much more power during the summer than you are in the winter, this may not be an issue.

More importantly, compared to leaving the panels on a flat surface, you are almost doubling the amount of power you can generate during the three bleakest months of the year. This means you can reduce the number of solar panels you need to install.

Optimum summer settings

If you wish to get the best output of your system overall, you will find that you will get slightly more energy, when measured over the course of the whole year, by angling your solar panels to an optimum summer time tilt.

In warm climates, where maximum energy consumption is during hot weather, angling your panels to get the maximum amount of sunlight during the height of summer can be the best solution, both financially and environmentally.

For Northern Hemisphere countries, the optimum summer time tilt is the optimum angle for May and July. For Southern Hemisphere countries, the optimum summer time tilt is the optimum angle for November and January:

90° – your latitude + 15.6° = optimum summer setting

Positioning your solar panels

Regardless of where you live, the sun always rises from the east and sets in the west. If you live in the Northern Hemisphere, solar panels will always work best if they are south-facing. In the Southern Hemisphere, solar panels work best if they are north-facing.

However, it is not always possible to position your solar panels so they are facing exactly the right way. For instance, if you want to install solar panels on the roof of the house, and the roof faces east/west, then it may not be practical to install solar panels in any other location.

Thanks to improved solar panel design over the past decade, this is not as big a problem as was once the case. Whilst the figure varies slightly in different parts of the world, from one solar panel manufacturer to another and during the different seasons of the year, the average efficiency drop of a solar panel mounted away from due south (due north in the Southern Hemisphere) is around 1.1% for every five degrees.

This means that if your panels face due east or due west, you can expect around 20% loss of efficiency compared to facing your panels in the optimum position.

You can even face your panels in the completely opposite direction – north in the Northern Hemisphere or south in the Southern Hemisphere – losing around 40% of the efficiency of your solar array by doing so.

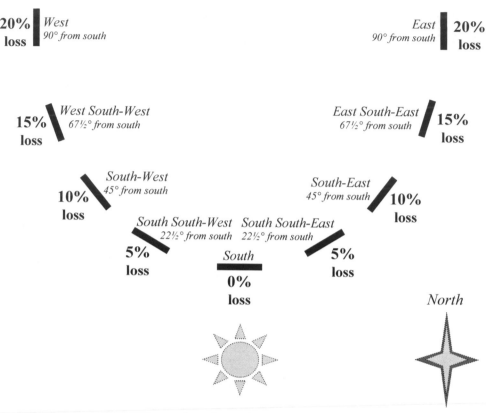

For the Northern Hemisphere, this chart shows the approximate efficiency loss by not facing your panels directly south. For Southern Hemisphere countries, the chart should be reversed.

Using solar irradiance to work out how much energy a solar panel will generate

Based on these figures, we can calculate on a monthly basis how much power a solar panel will give us per day, by multiplying the monthly solar irradiance figure by the stated wattage of the panel:

Solar Irradiance x Panel Wattage = Watt-hours per day

As we now know, the solar irradiance figure depends on the month and the angle for the solar panel. Assuming we have a 20-watt solar panel, mounted flat on the ground, here are the calculations for London in December and June:

	December	June
Flat	0.60 x 20W = 12 Wh of energy per day	4.86 x 20W = 97 Wh of energy per day

As you can see, there is a big difference in the amount of energy you can generate in the middle of summer, compared to winter. In the example above, over eight times the amount of energy is generated in the height of summer compared to the depths of winter.

Here are the same calculations again, but with the solar panels angled at 38° for best all-year-round performance. Note the significant improvement in winter performance and the slightly reduced summer performance:

	December	June
38° angle	1.05 x 20W = 21 Wh of energy per day	4.20 x 20W = 84 Wh of energy per day

Using solar irradiance to give you an approximate guide for the required power capacity of your solar array

In the same way that you can work out how much energy a solar panel will generate per day, you can use solar irradiance to give you an approximate guide for the required capacity of solar array that you need.

I say an approximate guide, because the *actual* capacity will also need to take into account:

- The peculiarities of your site
- The location and angles of your solar panels
- Any obstacles blocking the sunlight at different times of year

I cover all this in the next chapter when I look at the site survey.

Nevertheless, it can be useful to carry out this calculation in order to establish a ballpark cost for your solar electric system. The calculation is simple: take the figure you calculated for your total number of watt-hours per day and divide it by the solar irradiance figure for the worst month that you require your system to work.

Using our holiday home as an example, we can look at our watt-hours per day figure of 695Wh/day and then divide this number by the worst month on our irradiance chart (December). It is worth doing this based on mounting the solar panel at different angles, to see how the performances compare:

Flat	$695 \div 0.6 = 1159$ watts	If we have our solar panels laid flat, we would need a 1,159-watt solar array to power our home in December.
Upright	$695 \div 1.01 = 688$ watts	If we mount the solar panels vertically against a wall, we could generate the same amount of power with a 688-watt solar array.
38° angle Best year-round tilt	$695 \div 1.05 = 661$ watts	Angled towards the equator, we could generate the same amount of power with a 661-watt solar array.
23° angle Best winter tilt	$695 \div 1.08 = 643$ watts	With the optimum winter tilt, we can use a 643-watt solar array.
53° angle Best summer tilt	$695 \div 0.97 = 716$ watts	Angled towards the summer sun, we would require a 716-watt solar array to provide power in December
Tilt adjusted each month	$695 \div 1.08 = 643$ watts	With the tilt of the solar panel adjusted each month, we can use a 643-watt solar array, the same as the best winter tilt settings.

This chart tells us that to provide full power for our holiday home in December, we require a solar array with a generation capacity of between 643 watts and 1159 watts, depending on the tilt of the solar panels.

But remember our scope. We only want to use the home full time from March to October. The solar electric system only needs to provide enough electricity for a long weekend during the winter.

This means that, so long as our batteries are big enough to provide electrical power for a few days, it does not matter if the solar power in winter is not enough to provide for constant use. As soon as we close up the holiday home again, the solar panels will recharge the batteries.

Here are my calculations again, this time using October as our worst month:

Flat	$695 \div 1.69 = 411$ watts	If we have our solar panels laid flat, we would need a 411-watt solar array to power our home in October.

 Upright	$695 \div 2.07 = 335$ watts	If we mount the solar panels vertically against a wall, we could generate the same amount of power with a 335-watt solar array.
 38° angle **Best year-round tilt**	$695 \div 2.41 = 288$ watts	Angled towards the equator, we could generate the same amount of power with a 288-watt solar array.
 23° angle **Best winter tilt**	$695 \div 2.37 = 293$ watts	With the optimum winter tilt, we can use a 293-watt solar array.
 53° angle **Best summer tilt**	$695 \div 2.33 = 298$ watts	Angled towards the summer sun, we would require a 298-watt solar array to provide power in October
 Tilt adjusted **each month**	$695 \div 2.41 = 288$ watts	With the tilt of the solar panel adjusted each month, we can use a 288-watt solar array, the same as the best year-round tilt settings.

This chart tells us that we require a solar array with a generation capacity of between 288 watts and 411 watts, depending on the tilt of the solar panels. Compared to our earlier calculations for generating power throughout the year, it is much lower. We have just saved ourselves a significant amount of money.

You can also see that during the summer months, the solar electric system will generate considerably *more* electricity than we will need to run our holiday home. That is fine. Too much is better than not enough and it allows for the occasions when a light is left switched on or a TV is left on standby.

Solar panels and shade

The biggest negative impact on solar energy production is shade. Even if only a very small amount of your solar array is in shade, the impact on the performance of your whole system can have a very big effect.

Unlike solar thermal (hot water) systems, the loss of power through shading is much greater than the amount of the panel that is in shade. With solar thermal systems, if 5% of the panel is in shade, you lose around 5% of the power production.

Depending on the exact circumstances, even if only 5% of a photovoltaic solar panel is in shade, it is possible to lose 50–80% of power production from your entire solar array.

For this reason, it is hugely important that your solar energy system remains out of shade throughout the day. Sometimes this is not possible, and this requires some additional design work in order to keep the effect of shade on your system to a minimum.

I cover shade in much more detail in Appendix A, including an explanation of why shade has such a big impact on energy production. For now, it is just important to know that shading can significantly affect the amount of energy you can get from your solar energy system.

Solar array power point efficiencies

Now we know the theoretical size of our solar panels. However, we have not taken into account the efficiencies of the panels themselves or the efficiencies of the controller or inverter that handles them.

Solar panels are rated on their 'peak power output'. Peak power on a solar panel in bright sunlight is normally generated at between 14 and 20 volts. This voltage can go up and down quite significantly, depending on the amount of sunlight available.

This swing in voltage gets much higher if you have multiple solar panels connected together in series – or if you are using a higher-voltage solar panel. It is common for a solar array with fifteen or twenty panels connected in series to have voltage swings of several hundred volts when a cloud obscures the sun for a few seconds!

Managing this voltage swing can be done in one of two ways. The cheap and simple method is to cut the voltage down to a setting that the solar panel can easily maintain. For instance, a solar panel rated at 12 volts will usually maintain a voltage level of at least 14 volts during daylight hours. A charge controller or inverter that cuts the voltage down to this level will then always be able to use this power. The disadvantage of this approach is that, as you cut the voltage, the wattage drops with it, meaning you can lose a significant amount of energy.

In terms of the amount of energy you can capture, as opposed to what the solar array collects, this method can reduce the efficiency of a solar array by around 25%.

A better solution is to use controllers and inverters that incorporate *Maximum Power Point Tracking* (MPPT). Maximum power point tracking adjusts the voltage from the solar array to provide the correct voltage for the batteries or for the inverter in order to remove this inefficiency.

Maximum power point trackers are typically 90–95% efficient. Over the past three years, the price of MPPT controllers and inverters has dropped and the availability has increased to the point where it is almost always worth buying an MPPT controller and inverter for all but the very smallest and simplest solar installations.

Incidentally, you only need an MPPT inverter for grid-tie systems where you are powering the inverter directly from the solar panels. You do not require an MPPT inverter if you are planning to run the inverter through a battery bank.

To take into account power point efficiencies, you need to divide your calculation by 0.9 if you plan to use a MPPT controller or inverter, and by 0.75 if you plan to use a non-MPPT controller or inverter:

Non MPPT controller/inverter calculation		MPPT controller/inverter calculations	
Flat solar panel	Solar panel at 38° tilt	Flat solar panel	Solar panel at 38° tilt
411 watts ÷ 0.75 = **548 watt solar panel**	288 watts ÷ 0.75 = **384 watt solar panel**	411 watts ÷ 0.9 = **456 watt solar panel**	288 watts ÷ 0.9 = **320 watt solar panel**

The effects of temperature on solar panels

Solar panels will generate less power when exposed to high temperatures compared to when they are in a cooler climate. Solar PV systems can often generate more electricity on a day with a cool wind and a hazy sun than when the sun is blazing and the temperature is high.

When solar panels are given a wattage rating, they are tested at 25°C (77°F) against a 1,000 W/m² light source. At a cooler temperature, the solar panel will generate more electricity, whilst at a warmer temperature the same solar panel will generate less.

As solar panels are exposed to the sun, they heat up, mainly due to the infrared radiation they are absorbing. As solar panels are dark, they can heat up quite considerably. In a hot climate, a solar panel can quite easily heat up to 80–90°C (160–175°F).

Solar panel manufacturers provide information to show the effects of temperature on their panels. Called a *temperature coefficient of power* rating, it is shown as a percentage of total power reduction per 1°C increase in temperature.

Typically, this figure will be in the region of 0.5%, which means that for every 1°C increase in temperature, you will lose 0.5% efficiency from your solar array, whilst for every 1°C decrease in temperature you will improve the efficiency of your solar array by 0.5%.

Assuming a temperature coefficient of power rating of 0.5%, this is the impact on performance for a 100W solar panel at different temperatures:

	5°c / 41°F	15°c / 59°F	25°c / 77°F	35°c / 95°F	45°c / 113°F	55°c / 131°F	65°c / 149°F	75°c / 167°F	85°c / 185°F
Panel output for a 100W solar panel	110W	105W	100W	95W	90W	85W	80W	75W	70W
Percentage gain/loss	10%	5%	0%	-5%	-10%	-15%	-20%	-25%	-30%

In northern Europe and Canada, high temperature is not a significant factor when designing a solar system. However, in southern states of America and in Africa, India, Australia and the Middle East, where temperatures are significantly above 25°C (77°F) for much of the year, the temperature of the solar panels may be an important factor when planning your system.

If you are designing a system for all-year-round use, then in all fairness a slight dip in performance at the height of summer is probably not an issue for you. If that is the case, you do not need to consider temperature within the design of your system and you can skip the information on the next page.

If your ambient temperature is high during the times of year you need to get maximum performance from your solar panels, then you will need to account for temperature in your design.

You can help reduce the temperature of your solar panels by ensuring a free flow of air both above and below the panels. If you are planning to mount your solar panels on a roof, make sure there is a gap of around 7–10cm (3–4") below the panels, to allow a flow of air around them. Alternatively, you can consider mounting the panels on a pole, which will also aid cooling.

For a roof-top installation, if the average air temperature at a particular time of year is 25°C/77°F or above, multiply this temperature *in Celsius* by 1.4 in order to get a likely solar panel temperature. For a pole-mounted installation, multiply your air temperature by 1.2 in order to calculate the likely solar panel temperature. Then increase your wattage requirements by the percentage loss shown in the *temperature coefficient of power* rating shown on your solar panels, in order to work out the wattage you need your solar panels to generate.

Temperature impact on solar performance in Austin, Texas during the summer months

By way of an example, here is a table for Austin in Texas. This shows average air temperatures for each month of the year, the estimated solar panel temperature for the hottest months of the year and the impact on the performance on the solar array, assuming a temperature coefficient of power rating of 0.5%.

	Jan	Feb	Mar	Apr	May	Jun	Jul	Aug	Sep	Oct	Nov	Dec
Average monthly temperature	49°F 10°C	53°F 12°C	62°F 16°C	70°F 21°C	76°F 24°C	81°F 27°C	85°F 29°C	85°F 29°C	81°F 27°C	71°F 22°C	61°F 16°C	52°F 11°C
Likely roof-mounted temperature of solar array *(Celsius x 1.4)*					93°F 34°C	100°F 38°C	106°F 41°C	106°F 41°C	100°F 34°C			
Performance impact for roof-mounted solar:					-4½%	-6½%	-8%	-8%	-6½%			
Likely pole-mounted temperature of solar array *(Celsius x 1.2)*					84°F 29°C	90°F 32°C	95°F 35°C	95°F 35°C	90°F 32°C			
Performance impact for pole-mounted solar:					-2%	-3½%	-5%	-5%	-3½%			

The *Performance Impact* calculations in rows 3 and 5 of the above table are calculated using the following formula:

(Estimated Solar Panel Temp °C – 25°C) x (–Temp Coefficient of Power Rating)

So, for July, the calculation for roof-mounted solar was (41°C – 25°C) x (–0.5) = –8%

For around five months of the year, the ambient temperature in Austin is greater than 25°C/77°F. During these months, the higher temperature will mean lower power output from a solar array. If you are designing a system that must operate at maximum efficiency during the height of summer, you will need to increase the size of your solar array by the percentages shown, in order to handle this performance decrease.

You can find the average ambient air temperature for your location by visiting *The Weather Channel* website at *www.weather.com*. This excellent site provides average monthly temperatures for towns and cities across the world, shown in your choice of Fahrenheit or Celsius.

Our example holiday home project is in the United Kingdom, where the temperature is below 25°C for most of the year. In addition, our solar design will produce more power than we need during the summer months. As a result, we can ignore temperature in our particular project.

Working out an approximate cost

It is worth stressing again that these figures are only approximate at this stage. We have not yet taken into account the site itself and we are assuming that shading will not be an issue.

If you are planning to do the physical installation yourself, a solar electric system consisting of solar array, controller and battery costs around £3.50 ($5.60 US) per watt, +/– 15%.

A grid-tie system is cheaper to install than a stand-alone system, as you do not need to budget for batteries. You will, however, need a qualified electrician to certify the system before use. In most countries, you will also need all the components used in your solar installation to be certified as suitable for grid installation. If you are planning grid-tie, budget around £2.00–£3.00 ($3.10–$4.60 US) per watt, +/– 15%.

For our holiday home installation, we need 320 watts of solar electricity if we tilt the solar panels towards the sun, or 456 watts if we mount the panels flat. Our rough estimate suggests a total system cost of around £1,120 ($1,780 US), +/– 15% for tilted panels, or £1,596 ($2,540 US) +/– 15% for a flat panel installation.

If you remember, the cost to connect this holiday home to a conventional electricity supply was £4,500 ($7,200). Therefore, installing solar energy is the cheaper option for providing electricity for our home.

What if the figures do not add up?

In some installations, you will get to this stage and find out that a solar electric system simply is unaffordable. This is not uncommon. I was once asked to calculate the viability for using 100% solar energy at an industrial unit once, and came up with a ballpark figure of £33½m (around $54m)!

When this happens, you can do one of two things: walk away, or go back to your original scope and see what can be changed.

The best thing to do is go back to the original power analysis and see what savings you can make. Look at the efficiencies of the equipment you are using and see if you can make savings by using lower-energy equipment or changing the way equipment is used.

If you are absolutely determined to implement a solar electric system, there is usually a way to do it. However, you may need to be ruthless as to what you have to leave out.

In the example of the industrial unit, the underlying requirement was to provide emergency lighting and power for a cluster of computer servers if there was a power cut. The cost for implementing this system was around £32,500 ($52,000): comparable in cost to installing and maintaining on-site emergency generators and UPS equipment.

Working out dimensions

Now we know the capacity of the solar panels, we can work out an approximate size of our solar array. This is extremely useful information to know before we carry out our site survey: the solar panels have to go somewhere. We need to be able to find enough suitable space for them where they will receive uninterrupted sunshine in a safe location.

There are two main technologies of solar panels on the market: *amorphous* and *crystalline* solar panels. I will explain the characteristics and the advantages and disadvantages of each, later on.

For the purposes of working out how much space you're going to need to fit the solar panels, you need to know that a 1m² (approximately 9.9 square feet) amorphous solar panel generates in the region of 60 watts, whilst a 1m² crystalline solar panel generates in the region of 160 watts.

Therefore, for our holiday home, we are looking for a location where we can fit between 5 and 7.6m² (49–75 square feet) of amorphous solar panels or 2–3m² (20–29 square feet) of crystalline solar panels.

In conclusion

- By calculating the amount of solar energy *theoretically* available at our site, we can calculate ballpark costs for our solar electric system
- There are various inefficiencies that must be considered when planning your system. If you do not take these into account, your system may not generate enough power
- It is not unusual for these ballpark costs to be far too expensive on your first attempt. The answer is to look closely at your original scope and see what can be cut in order to produce a cost-effective solution
- As well as working out ballpark costs, these calculations also help us work out the approximate dimensions of the solar array. This means we know how much space we need to find when we are carrying out a site survey

Surveying Your Site

The site survey is one of the most important aspects of designing a successful solar system. It will identify whether or not your site is suitable for solar. If it is, the survey identifies the ideal position to install your system, ensuring that you get the best value for money and the best possible performance.

What we want to achieve

For a solar electric system to work well, we need the site survey to answer two questions:

- Is there anywhere on the site that is suitable for positioning my solar array?
- Do nearby obstacles such as trees and buildings shade out too much sunlight?

The first question might at first sound daft but, depending on your project, it can make the difference between a solar energy system being viable or not.

By answering the second question, you can identify how much of the available sunlight you will be able to capture. It is vitally important that you answer this question. The number one reason for solar energy failing to reach expectations is obstacles blocking out sunlight, which dramatically reduces the efficiency of the system.

To answer this second question, we need to be able to plot the position of the sun through the sky at different times of the year. During the winter, the sun is much lower in the sky than it is during the summer months. It is important to ensure that the solar array can receive direct sunlight throughout the day during the winter.

What you will need

You will need a compass, a protractor, a spirit level and a tape measure.

Inevitably a ladder is required if you are planning to mount the solar array on a roof. A camera can also be extremely useful for photographing the site. If you have an iPhone or an Android cell phone, you can also download some cheap software that will help you identify the path of the sun across the sky and assist with obstacle analysis.

I also find it useful to get some large cardboard boxes. Open them out and cut them into the rough size of your proposed solar array. This can help you when finding a location for your solar panels. It is far easier to envisage what the installation will look like and it can help highlight any installation issues that you might otherwise have missed.

If you have never done a solar site survey before, it does help if you visit the site on a sunny day.

Once you have more experience with doing solar site surveys, you will find it does not actually make much difference whether you do your site survey on a sunny day or an overcast day. As part of the site survey, we manually plot the sun's position across the sky, so once you have more experience, sunny weather actually makes little difference to the quality of the survey.

First impressions

When you first arrive on the site, the first thing to check is that the layout of the site gives it access to sunlight.

We will use a more scientific approach for checking for shade later, but a quick look first often highlights problems without needing to carry out a more in-depth survey.

If you are in the Northern Hemisphere, look from east, through south and to the west to ensure there are no obvious obstructions that can block the sunshine, such as trees and other buildings. If you are in the Southern Hemisphere, you need to check from east, through north to west for obstructions. If you are standing on the equator, the sun passes overhead, so only obstructions in the east and west are important.

Be very careful not to look directly at the sun, even for a few moments, whilst you are carrying out this survey. Even in the middle of winter, retina burn can cause permanent damage to your eyesight.

Look around the site and identify all the different options for positioning the solar array. If you are considering mounting your solar array on a roof, remember that the world looks a very different place from a roof-top, and obstructions that are a problem standing on the ground look very different when you are at roof height.

Drawing a rough sketch of the site

It can be helpful to draw up a rough sketch of the site. It does not have to be accurate, but it can be a useful tool to have, both during the site survey and afterwards when you are designing your system.

Include all properties and trees that are close to your site and not just those on your land. Include trees that are too small to worry about now, but may become a problem in a few years' time. Also make a note of which way is north.

Positioning the solar array

Your next task is to identify the best location to position your solar array. Whilst you may already have a good idea where you want to install your solar panels, it is always a good idea to consider all the different options available to you.

As we discovered in the last chapter, solar arrays perform at their best when tilted towards the sun.

If you are planning to install solar energy for a building, then the roof of the building can often be a suitable place to install the solar array. This is effective where the roof is south-facing or where the roof is flat and you can fit the panels using angled mountings.

Other alternatives are to mount solar panels on a wall. This can work well with longer, slimmer panels that can be mounted at an angle without protruding too far out from the wall itself. Alternatively, solar panels can be ground-mounted or mounted on a pole.

When considering a position for your solar array, you need to consider how easy it is going to be to clean the solar panels. Solar panels do not need to be spotless, but dirt and grime will reduce the efficiency of your solar system over time, so while you are looking at mounting solutions it is definitely worth considering how you can access your panels to give them a quick wash every few months.

Roof-mounting

If you are planning to mount your solar array on a roof, you need to gain access to the roof to check its suitability.

Use a compass to check the direction of the roof. If it is not directly facing south, you may need to construct an angled support in order to get the panels angled correctly.

You will also need to find out the pitch of the roof. Professionals use a tool called a *Roof Angle Finder* to calculate this. Roof angle finders (also called *Magnetic Polycast Protractors*) are low-cost tools available from builders' merchants. You press the angle finder up against the rafters underneath the roof and the angle finder will show the angle in degrees.

Alternatively, you can calculate the angle using a protractor at the base of a roof rafter underneath the roof itself.

Solar panels in themselves are not heavy: 15–20 kilograms (33–44 pounds) at most. Yet when multiple panels are combined with a frame, especially if that frame is angled, the overall weight can become quite significant.

Check the structure of the roof. Ensure that it is strong enough to take the solar array and to ascertain what fixings you will need. It is difficult to provide general advice on roof structures and fixings. There are so many different roof designs it is not possible for me to provide much useful information on this subject. If you are not certain about the suitability of your roof, ask a builder or an architect to assess your roof for you.

Roof-mounting kits are available from solar panel suppliers. Alternatively, you can make your own.

If it does not compromise your solar design, it can be quite useful to mount your solar panels at the lowest part of the roof. This can make it considerably easier to keep the panels clean: most window cleaners will happily wash easily-accessible solar panels if they are situated at the bottom of the roof, and telescopic window cleaner kits are available to reach solar panels at the lower end of a roof structure.

Measure and record the overall roof-space available for a solar array. It is also a good idea to use the cardboard cut-outs you made earlier and place these on the roof to give a 'look and feel' for the installation and help you identify any installation issues you may have with positioning and mounting the solar array.

Ground-mounting

If you want to mount your solar array on the ground, you will need a frame onto which you can mount your solar panels. Most solar panel suppliers can supply suitable frames or you can fabricate your own on site.

There are benefits for a ground-mounted solar array: you can easily keep the panels clean and you can use a frame to change the angle of the array at different times of year to track the height of the sun in the sky.

Take a note of ground conditions, as you will need to build foundations for your frame.

Incidentally, you can buy ground-mounted solar frames that can also move the panel to track the sun across the sky during the day. These *Solar Trackers* can increase the amount of sunlight captured by around 15–20% in winter and up to 55% in summer.

Unfortunately, at present, commercial solar trackers are expensive. Unless space is at an absolute premium, you would be better to spend your money on a bigger solar array.

However, for a keen DIY engineer who likes the idea of a challenge, a solar tracker that moves the array to face the sun as it moves across the sky during the day could be a useful and interesting project to do. There are various sites on the internet, such as *instructables.com*, where keen hobbyists have built their own solar trackers and provide instructions on how to make them.

Pole-mounting

Another option for mounting a solar panel is to affix one on a pole. Because of the weight and size of the solar panel, you will need an extremely good foundation and a heavyweight pole, in order to withstand the wind.

You can mount up to 600-watt arrays using single-pole mountings. Larger arrays can be pole-mounted using frames constructed using two or four poles.

Most suppliers of solar panels and associated equipment can provide suitable poles.

Splitting the solar array into several smaller arrays

It may be that when you get to the site, you find that there is no one space available that will allow you to install all the solar panels you need. If this is the case, it is possible to split your single solar array into several smaller arrays. This means, for instance, that you could have two sets of panels mounted on different roof pitches, or some mounted on a roof and some from a pole.

If you do this, you are creating two *separate* solar energy systems, which you then have to link together. For a grid-tie system, you would require either a micro-inverter system or an inverter that can accept inputs from more than one solar array. For a stand-alone system, you would require a battery controller for each separate solar array.

Identifying the path of the sun across the sky

Once you have identified a suitable position for your solar array, it is time to be a little more scientific in ensuring there are no obstructions that will block sunlight at different times of the day, or at certain times of the year.

The path of the sun across the sky changes throughout the year. This is why carrying out a site survey is so important: you cannot just check to see where the sun is shining today. The height and position of the sun constantly changes throughout the year.

Each year, there are two days in the year when the day is exactly twelve hours long. These two days are 21st March and 21st September, the *solar equinoxes*. On these equinoxes the sun rises due east of the equator and sets due west of the equator. At solar noon on the equinox, exactly six hours after the sun has risen, the angle of the sun is 90° minus the local latitude.

In the Northern Hemisphere (i.e. north of the equator), the longest day of the year is 21st June and the shortest day of the year is 21st December. These two days are the summer and winter solstices respectively.

On the summer solstice, the angle of the sun is 23.5° higher than it is on the equinox, whilst on the winter solstice the angle is 23.5° lower than on the equinox.

These two extremes are due to the tilt of the earth, relative to its orbit around the sun. In the Northern Hemisphere, the summer solstice occurs when the North Pole is tilted towards the sun, and the winter solstice occurs when the North Pole is tilted away from the sun.

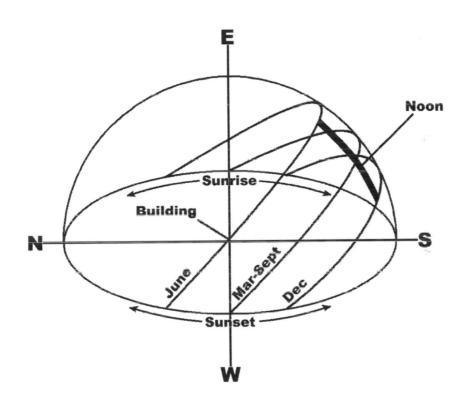

Figure 4: This chart shows the different paths of the sun from sunrise to sunset at different times of the year from the Northern Hemisphere. The intersection between N, S, E and W is your location.

We will take London in the United Kingdom as an example. London's latitude is 51°. On the equinox, the angle of the sun at noon will be 39° (90° − 51°). On the summer solstice, the angle will be 62.5° (39° + 23.5°) and on the winter solstice, it will be 15.5° (39° − 23.5°).

London in mid-summer   *London in mid-winter*

As well as the solar irradiance figures, Appendix B shows the height of the sun in the sky at noon at different times of the year and the optimum tilts for solar panels for the United States, Canada, Australia, New Zealand, United Kingdom and Ireland.

For more detailed information on sun heights on a monthly basis, or for information for other countries, visit *www.SolarElectricityHandbook.com* and follow the link to the solar panel angle calculator.

Shading

As mentioned in a previous chapter, shading can have a very significant impact on the performance of your solar energy system. Even a tiny amount of shading can have a huge impact on the amount of energy that your system is able to generate. Therefore, it is important that your solar array remains shade-free whenever possible.

You can carry out this analysis in various ways. You can use a professional *obstacle analysis* tool, you can download an obstacle analysis tool onto your cell phone or you can use paper and pencil to come up with a rough plan.

Professional tools for obstacle analysis

In the past, a product called a *Solar Pathfinder* was one of the best tools you could get. This was a plastic unit with an angle chart mounted on the top. A glass dome was then placed on top of the chart. You would mount the unit onto a tripod at the desired location. Obstacles were reflected in the glass bubble and this would allow you to manually plot the obstacles on the chart and then manually work out your shading issues.

The Solar Pathfinder is surprisingly effective, as it can be easily moved around in order to find the best location for solar panels. Many professional solar installers still use a Solar

Pathfinder for quick checks, despite also using the more expensive and advanced electronic equipment to provide a more detailed analysis.

Some solar suppliers can rent you a Solar Pathfinder for a small daily or weekly fee and can do the manual calculations for you once you have plotted the obstacles.

Today there are electronic systems that use GPS, tilt switches and accelerometers to do this work electronically. They are expensive to buy or rent on a daily basis. They do provide extremely comprehensive solar analysis, however, and if you plan to take up solar installations professionally, they are a worthwhile investment.

The best known is the *Asset* from Wiley Electronics and the *SunEye* from Solmetric. My personal preference is *SunEye*, as I find the unit simpler, but both products do a similar job.

Cell phone applications

Whilst they are very good, these tools are overkill for smaller solar installations and are often very complicated to use. Thankfully, modern smart phones provide the processing power and functionality to do a similar job. A few companies have now developed solar shade analysis software to run on these mobile phones. These use the phone's built-in GPS, compass, accelerometer and camera to record a complete shade analysis in a matter of a few moments.

If you have an iPhone, you can download a product called *Solmetric IPV*. Costing $29.99, this application handles your obstacle tracking, automatically providing charts showing your shading analysis throughout the year. The detail of reporting is not as great as some of the other electronic tools, but more and more professionals are now using this software. It provides most of the functions that you get with a more expensive system, but in a package that is easier to use and far cheaper to buy.

You can find out more about *Solmetric IPV* from *www.solmetric.com* and the software can be downloaded from the iTunes website.

Alternatively, there is an Android phone download called *Solar Shading*, produced by Comoving Magnetics. Costing $15 on the Android Market, this application provides you with a complete shading analysis throughout the year, presenting the information in easy-to-read charts.

Above: Examples of on-screen reports from Solmetric IPV running on the iPhone.
Below: Screen-shots from Solar Shading, running on Android phones.

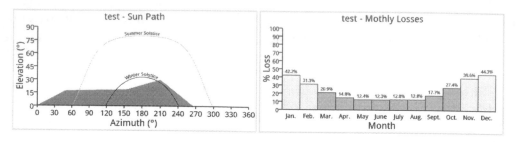

Using paper and pencil

Finally, if you do not have access to a professional tool or a suitable mobile phone, you can use the old-fashioned method of paper and pencil. Go to the position where you are planning to put your solar array and find due south with a compass. Looking from the same height as your proposed location, and working from east to west, check there are no obstacles, such as trees or buildings, that can shade the sun at different times of the day or when the sun is at its lowest winter height.

To do this, you will need to find out what position the sun rises and sets at different times of the year. Thankfully, this is easy to find out. The solar angle calculator which you can find at *www.SolarElectricityHandbook.com* includes this information, making it easy to identify the path of the sun at these different times of the year.

For a simple one-off site survey, the easiest way to identify potential obstructions is to use a protractor and pencil. Tape the pencil to the centre of the protractor where all the lines meet, in such a way that the other end of the pencil can be moved across the protractor. You can use this protractor to check the field of view, using the pencil as an 'aimer' to show the angle of the sun in the sky based on different times of the year.

Be very careful not to look directly at the sun, even for a few moments, whilst you are carrying out this survey. Even in the middle of winter, retina burn can cause permanent damage to your eyesight.

Your survey needs to ensure there are no obstacles in the depths of winter, when the sun is only a few degrees up in the sky.

In the case of London, on the 21st December the sun will be only 15½° high at midday (at due south) and lower than that for the rest of the day.

If there are obstacles that are blocking visibility of the sun, find another location. Alternatively, find other ways around the obstacle, such as mounting the solar array higher up on a frame.

Of course, if you do not need your solar system to produce much power during the winter months this may not be a problem for you. However, make sure obstacles do not shade your system for the times of year when you need power.

Future proof your system

You do need to consider the future when installing a solar electric system. The system will have a lifetime of at least 20 years, so as far as possible you need to ensure that the system will be effective for that length of time.

When scanning the horizon, take into account that trees and hedges will grow during the lifetime of the system. A small spruce in a nearby garden could grow into a monster in the space of a few years, and if that is a risk, it is best to know about it now, rather than have a nasty surprise a few years down the line.

See if there is any planned building work nearby that may overshadow your site and try to assess the likelihood of future building work that could have an impact on shading.

It is also worth finding out if fog or heavy mist is a problem at certain times of the year. The efficiency of your solar array will be compromised if the site has regular problems with heavy mist.

What if there are shading obstructions?

If there are obstructions that shade your proposed location, and there are no other locations that are suitable for solar, you need to ascertain at what point during the day the obstructions occur.

Anything due south (or due north in the Southern Hemisphere) is a major problem, as this will be the position of the sun when the intensity of the sunlight is at its highest. Core power generation occurs between 9am and 3pm. If you have shading either before 9am or after 3pm, you will lose around 20% of your capability in the summer, or 40% of your capability if you have shading both before 9am and after 3pm.

During the winter, the difference is not so great. If you have shading before 9am or after 3pm during the winter, you will probably lose only around 5–10% of your generating capabilities during this time.

If you have shading during your core power generation times, you need to give serious thought as to whether you should continue with a solar implementation: the performance of your solar system will be severely compromised.

If you do have significant shading issues and you want to find out the exact impact of these obstructions on your solar array, you are going to need to use a professional tool for obstacle analysis. The electronic tools will be able to quantify exactly what the impact of the shade is on your system at different times of the year.

If obstructions occur for part of the day, such as during the morning or during the afternoon, you can consider increasing the number of solar panels you purchase and angling them away from the obstruction in order to increase their collection of sunlight during the unobstructed parts of the day.

Alternatively, you may be better off investigating other energy options, such as wind power or fuel cells, either instead of using solar or in combination with a smaller solar electric system.

Appendix A looks at the issue of shading in a lot more detail and explains how you can work around obstruction issues. It is always best to avoid shade as far as possible, because workarounds can become costly. However, if you cannot eliminate shade altogether, it need not write off solar as a solution that can work for you.

Positioning batteries, controllers and inverters

You need to identify a suitable location for batteries. This could be a room within a building, in a garage or garden shed, or in a weatherproof battery housing.

It is important to try to keep all the hardware close together, in order to keep the cable lengths as short as possible. By 'hardware', I am referring to the solar array itself, batteries, controller and inverter.

For the batteries, inverter and controller, you are looking for a location that fits the following criteria:

- Water- and weather proof
- Not affected by direct sunlight
- Insulated to protect against extremes of temperature
- Facilities to ventilate gases from the batteries
- Protected from sources of ignition
- Away from children, pets and rodents

Lead acid batteries give off very small quantities of hydrogen when charging. Hydrogen is explosive. You must ensure that, wherever your batteries are stored, the area receives adequate external ventilation to ensure these gases cannot build up.

Because of the extremely high potential currents involved with lead acid batteries, the batteries must be in a secure area away from children and pets.

For all of the above reasons, batteries are often mounted on heavy-duty racking, which is then made secure using an open-mesh cage. Alternatively, you can purchase purpose-built battery enclosures from solar or battery suppliers.

Controllers and inverters need to be mounted as close to the batteries as possible. These are often wall-mounted, but can also be mounted to racking.

Large inverters can be extremely heavy, so if you are planning to wall-mount one, make sure that the wall is load-bearing and able to take the weight.

Cabling

While you are on site, consider likely routes for cables, especially the heavy-duty cables that link the solar array, controller, batteries and inverter together. Try to keep cable lengths as short as possible, as longer cables mean lower efficiency. Measure the lengths of these cables so that you can ascertain the correct specification for cables when you start planning the installation.

Site survey for the holiday home

Back to our holiday home example: based on our previous calculations, our holiday home needs a solar array capable of generating 320 watt-hours of energy, if we angle the array towards the sun. This solar array will take up approximately 2m² (18 sq. ft) of surface area.

Our site survey for the holiday home showed the main pitch of the roof is facing east to west. This is not ideal for a solar array. The eastern side of the roof has a chimney. There is shade from a tall tree that obscures part of the west-facing part of the roof. There is no space on the gable end of the roof to fit the required solar panels.

Road

Map of the holiday home, identifying likely obstacles and a suitable position for the solar array

An old shed close to the house has a south-facing roof, but only a 20° pitch. A tree shades the shed for most of the mornings for nine months of the year. Furthermore, the condition of the shed means that it would need significant remedial work should we decide to use it.

There is a farm to the east of the house, with a large barn and a number of trees bordering the house, the tallest of which is approximately 30 feet (10m) tall. One of these trees provides shade to the shed and part of the rear of the house during the winter, and may provide more shade during the rest of the year if it continues to grow.

The garden is south-facing and receives sunlight throughout the year.

We decide to install the solar array in the rear garden, constructing a suitable 1.2m (4 feet) tall garden store with a south-facing pitched roof of approximately 52° (allowing us to tilt the solar array at 38° from vertical for best year-round sunshine). Our *solar house* will hold the batteries and controller and will have adequate ventilation to ensure that the small amounts of hydrogen generated by the batteries can escape. By building our own structure, we can install the solar array at the optimum tilt to capture as much sunlight as possible. This means our solar array is compact and keeps our costs to a minimum.

The solar house will be located around 10m (33 feet) from the house and shielded from the house by a new shrubbery.

The cable lengths between the solar array and the solar controller are approximately 2m (6½ feet). The cable length between the solar controller and the batteries is less than 1m (3 feet). The cable length between the solar house and the home is 12m (40 feet). There is a further 10m (33 feet) of cabling inside the house.

These longer cable lengths are not ideal. Cable runs should be as short as possible in order to reduce power losses through the cable. However, we cannot position the solar array any nearer to the house. We will have to address this particular problem through our design.

In conclusion

- There is a lot to do on a site survey. It is important. Spend time; get it right.
- Drawing a map and taking photographs can help with the site survey and are invaluable for the next stage, when we start designing our new system.
- Solar panels can be roof-mounted, mounted with a frame on the ground or on a pole.
- Once you have identified a location for the panels, check for obstructions that will shade the panels throughout the year.
- These obstructions are most likely to be an issue during the winter months, when solar energy is at a premium.
- Identify a suitable space for batteries, controller and inverter.
- Plan the cable runs and the measure the length of the required cables.
- Cables should be as short as possible, in order to reduce the voltage losses through the system. If long cable lengths are necessitated by the positioning of the solar array, we may need to run our system at a higher voltage to compensate.

Understanding the Components

Once you have completed your site survey, you know all the facts: how much power you need to generate, the suitability of your site and approximately how much it is going to cost you.

Now you need to look at the different technologies and products that are available, to see what best suits you and your application.

Your choice of components and the design of your system will depend on whether you are designing a stand-alone system (which also includes grid fallback and grid failover systems), or a grid-tie system that exports energy to the grid.

Because there are differences in the design of stand-alone and grid-tie systems, I have split this section into four chapters. This chapter looks at components that are common to both grid-tie systems and stand-alone systems. The next chapter looks specifically at what you require for grid-tie systems. The third chapter looks specifically at stand-alone systems and systems that incorporate their own battery store, whilst the fourth chapter looks at the component certification and regulations that you need to take into account when selecting solar energy equipment.

How to use these chapters

These next three chapters will go into much more detail about the different options available to you. There is a bewildering choice of solar panels, batteries, controllers, inverters and cables.

These chapters explain the technology in a lot more detail, so you can go and talk sensibly to suppliers and understand what they are saying.

Common components for all systems

Core to all solar energy systems are the solar panels themselves. Most solar panels can be used for either grid-tie or stand-alone use, and although recently some manufacturers have released higher-voltage solar panels designed specifically for grid-tie applications, the criteria for choosing one solar panel over another remain the same.

Solar panels

There are three different technologies used for producing solar panels. Each has its own set of benefits and disadvantages.

For the purpose of this handbook, I am ignoring the expensive solar cells used on satellites and in research laboratories and focusing on the photovoltaic panels that are available commercially at reasonable cost today.

Amorphous solar panels

The cheapest solar technology is amorphous solar panels, also known as *thin-film* solar panels.

These panels have had a bad reputation in the past, with poor product reliability and questionable lifespan. This has often been down to the chemistries used in older designs of panel breaking down under extremes of temperature over a period of a few years, or the poor quality of materials used in the production of cheap panels.

Thankfully, this technology has matured significantly over the past five years and amorphous solar is now regarded as being highly reliable, with some significant benefits over traditional solar panels. Big name manufacturers such as Mitsubishi, Sanyo and Sharp now manufacture high-quality amorphous solar panels, along with some exceptionally good specialist manufacturers such as Solar Frontier and Uni-Solar. Some manufacturers now even offer a ten- or twenty-year warranty on their amorphous panels.

On paper, amorphous solar panels are the least-efficient panels available, typically converting around 6–8% of available sunlight to electricity. This means that you need twice as much space available for installing amorphous solar panels compared to crystalline panels.

However, amorphous panels are good at generating power even on overcast and extremely dull days. In general, they are also far better in extreme temperature conditions, with significantly less power loss at higher temperatures than other solar panel technologies.

Unlike other solar panel technologies, amorphous solar panels provide excellent performance even when partially shaded. Whilst a best-case scenario is to eliminate shading whenever and wherever possible, amorphous panels continue to operate at a high level of efficiency even if part of the array is in shade.

Amorphous panels can also be manufactured into a shape or mounted on a curved surface. They can be made to be hard-wearing enough to be fitted onto surfaces that can be walked on. A few solar manufacturers have started manufacturing amorphous solar roof tiles (or shingles), so that new-build houses can incorporate solar into the structure of the roof.

This combination makes amorphous panels suitable for integration into consumer products such as mobile phones and laptop computers, and for mobile products such as the roof of an RV or caravan, where the manufacturer has no control over where the products are placed or how they are used.

Amorphous panels are the cheapest panels to manufacture and a number of manufacturers are now screen-printing low-cost amorphous solar films. Between 2008 and 2013, amorphous solar panel costs dropped by around 30% each year. Whilst the prices are dropping more slowly now, the success of amorphous solar panels have been a big factor of driving down the costs of the entire solar industry.

Because of their lower efficiency, an amorphous solar panel has to be much larger than the equivalent polycrystalline solar panel. As a result, amorphous solar panels can only be used either where there is no size restriction on the solar array or where the overall power requirement is very low.

In terms of environmental impact, amorphous panels tend to have a much lower carbon footprint at point of production, compared to other solar panels. A typical carbon payback for an amorphous solar panel would be in the region of 12–30 months.

Most amorphous solar panels have comparatively low power outputs. These panels can work well for smaller installations of up to around 300-watt outputs, but not so well for larger installations: larger numbers of panels will be required and the additional expense in mounting and wiring these additional panels starts to outweigh their cost advantage.

Consequently, amorphous solar panels are often more suited to OEM applications, as an energy source built into a manufactured product, or for large-scale commercial installations where the panels are incorporated into the structure of a roof-space on a new build.

Some of the most exciting advances in solar technologies over the past three years have come from amorphous technology. Products as diverse as mobile phones, laptop computers, clothing and roofing materials have all had amorphous solar panels built into them. An exciting technology, amorphous solar is going to get better and better over the coming years.

Polycrystalline solar panels

Polycrystalline solar panels are made from multiple solar cells, each made from wafers of silicon crystals. They are far more efficient than amorphous solar panels in direct sunlight, with efficiency levels of 13–18%.

Consequently, polycrystalline solar panels are often around one third of the physical size of an equivalent amorphous panel, which can make them easier to fit in many installations.

Polycrystalline solar panels typically have a life expectancy of about 25 years. This can often be exceeded: commercial solar panels only became available in the late 1970s and early 1980s and many of these panels are still perfectly functional and in use to this day.

The manufacturing process for polycrystalline solar panels is complicated. As a result, polycrystalline solar panels are expensive to purchase, often costing 20–30% more than amorphous solar panels. The environmental impact of production is also higher than amorphous panels, with a typical carbon payback of 3–5 years.

Prices for polycrystalline solar panels are dropping, thanks to both the increase in manufacturing capacity over the past few years and the increasing popularity for larger screen televisions, which use the same specification glass. Between 2007 and 2013, prices dropped by around 25-30% per year. This past year, prices have stabilized. Most industry experts agree that prices are likely to remain at their current levels for the next 18-24 months.

Monocrystalline solar panels

Monocrystalline solar panels are made from multiple smaller solar cells, each made from a single wafer of silicon crystal. These are the most efficient solar panels available today, with efficiency levels of 15–24%.

Monocrystalline solar panels have the same characteristics as polycrystalline solar panels. Because of their efficiencies, they are the smallest solar panels (per watt) available.

Monocrystalline solar panels are the most expensive solar panels to manufacture and therefore to buy. They typically cost 35–50% more than the equivalent polycrystalline solar panels.

Which solar panel technology is best?

For most applications, polycrystalline panels offer the best solution, with reasonable value for money and a compact size.

Amorphous panels can be a good choice for smaller installations where space is not an issue. They are usually not practical for generating more than a few hundred watts of power because of their overall size, unless you have an extremely large area that you can cover with solar panels.

What to look for when choosing a solar panel

Not all solar panels are created equal, and it is worth buying a quality branded product over an unbranded one. Cheaper, unbranded solar panels may not live up to your expectations, especially when collecting energy on cloudy days.

If you are spending a lot of money buying a solar energy system that needs to last many years, it is advisable to purchase from a known brand such as Kyocera, Panasonic, Clear Skies, Hyundai, Sanyo, Mitsubishi, Solar Frontier or Sharp. My personal recommendation is Kyocera polycrystalline solar panels, or Mitsubishi or Solar Frontier amorphous panels. I have found these to be particularly good.

Buying cheap solar panels

Not all solar energy systems have to last ten or twenty years. If you are looking for a small, cheap system to provide power to an RV or caravan, or your requirements are modest, such as installing a light in a shed, buying a cheap solar panel may well be the right option for you.

The quality of the cheaper solar panels has improved significantly over the past few years. Six or seven years ago, buying a cheap, unbranded Chinese-made product was a recipe for disaster. Many of the panels were poorly assembled, allowing water to seep through the frames and damaging the solar cells. A lot of them used plate glass, often a thin, low-grade glass that becomes clouded over time and is easily chipped or broken. The cells used by these manufacturers were often sub-standard reject cells and often degraded very quickly.

Thankfully, most of these problems are now resolved and if you buy a cheap solar panel on eBay, you are likely to have a good product that will reliably generate power for five to ten years, and in all probability a lot longer. If you are buying a solar panel from a manufacturer you have never heard of, here is a checklist of things to look for:

Buy bigger than you think you'll need

If you are buying a very cheap solar panel, you are likely to be saving as much as 50% of the price when compared to buying a branded product. However, expect it to degrade slightly more quickly than a branded unit, and do not expect it to be quite so efficient.

To counter this, buy a solar panel with a higher watt rating (often shown as a watt peak, or Wp, rating) than you actually need, or buy additional solar panels if you are purchasing an entire array. Aim for 15% more power than you would otherwise have bought. You will still save a lot of money, but you will have an extra bit of assurance that the system will be up to the task.

Warranty

With cheap solar panels, you're not going to get a five-, ten- or twenty-year warranty, but you should still expect a one- or two-year warranty with any solar panel you buy. Check to see exactly what the warranty offers.

You are looking for a warranty that guarantees a minimum output under controlled conditions. The standard across the industry is to guarantee 80% of the quoted output under controlled conditions.

If you have a warranty claim, also check to see how you can claim on that warranty. Shipping a broken solar panel half way around the world and paying for return carriage is likely to cost as much as buying a new solar panel.

Glass

If you are buying a solar panel that is going to be fitted to a moving vehicle, or if you are buying a physically large solar panel, make sure that the solar panel uses tempered glass.

Tempered, or toughened, glass is around eight times stronger than plate glass. This makes it far more robust. If your glass is chipped on your solar panel, you will immediately see a significant drop in power output. If water gets into the solar panel itself, it can create a short circuit and becomes a fire hazard. Water and electricity do not mix.

It is worth noting that some amorphous solar panels cannot use tempered glass because of the way the thin film is applied to the glass. Some manufacturers reduce this problem by using thicker plate glass.

Second-hand solar PV panels

From time to time, second-hand solar panels appear for sale. They appear on eBay or are sold by solar equipment suppliers or building salvage yards.

So long as they come from a reputable brand, second-hand solar panels can be extremely good value for money and even old panels that are 25–30 years old may still give many more years of useful service. Although good quality solar panels should provide at least 25 years service, nobody knows how much longer they will last. The early commercially available solar panels (which are now over 30 years old) are still working extremely well, typically working at around 80–90% of their original capacity.

There are, however, a few points to look out for if you are considering buying second-hand solar PV panels:

- Never buy second-hand solar PV panels unseen. Take a multi-meter with you and test them outside to make sure you are getting a reasonable voltage and wattage reading
- Check the panels and reject any with chipped or broken glass. Also reject any panels where the solar cells themselves are peeling away from the glass or have condensation between the glass and the solar cell
- The efficiency of older solar PV panels is significantly lower than new panels. 30 years ago, the most efficient solar panels were only around 5–6% efficient, compared to 13–24% efficiency levels today. 10–15 years ago, most solar panels were around 10–12% efficient

- Consequently, a solar PV panel from the early 1980s is likely to be three times the size and weight of an equivalent modern crystalline panel
- These second-hand panels will not have any of the safety certification ratings that you get with new solar panels. This may cause issues with building regulations or building insurance, if you are installing these onto a building as part of a new solar installation.

Concentrated Photovoltaics and Solar Concentrators

Concentrated Photovoltaics (CPV) technology uses Fresnel lenses and curved mirrors to concentrate a large area of sunlight onto a smaller area of photovoltaic cells. They do this by refracting the light to concentrate it and increase the intensity of sunlight in a smaller area.

In effect, by concentrating the sunlight into a smaller area and increasing the solar irradiance, significantly more energy can be captured by the solar panel, thereby improving its efficiency quite impressively.

However, there are problems with this technology. Most specifically, the heat build-up is quite considerable and, in testing, many solar panels have been destroyed by the excessive heat generated by the concentrated sunlight. This is especially true of Fresnel lenses built by enthusiastic amateurs.

The panels tend to be quite large and bulky. Due to the heat build-up, they also need to be very carefully mounted, with adequate ventilation around the panel.

There are now a few solar concentrators on the market, but the technology has, by and large, being overtaken by events. The logic behind their design was that solar panels were expensive and lenses and mirrors were cheap. As the price of solar has dropped so much over the past few years, the economics for solar concentrators no longer add up.

A few installations use mirrors or polished metal to reflect additional sunlight back onto solar panels and therefore increasing the solar irradiance. However, you must take care to ensure that the reflected light does not dazzle anyone. The practicalities of mounting, safety and ensuring that people are not dazzled by the reflected sunlight normally dissuade people from using mirrors in this way.

Solar panel mountings

You can either fabricate your own mounting for your solar panels, or purchase a ready-made modular system.

The design of the system must take into account wind loading, so that it is not damaged or destroyed in high winds. If you are installing solar in a hot climate, your mounting must also ensure there is adequate ventilation behind the panel to avoid excessive heat build-up.

Your support structure needs to be able to set the angle of the solar array for optimal positioning towards the sun.

If you have not installed solar electric systems before, it is usually a good idea to buy a modular support structure from the same supplier as your solar panels. Once you have more experience, you can then choose to fabricate your own, if you prefer.

Solar trackers

For ground- or pole-mounted solar arrays, you can buy solar trackers that track the path of the sun across the sky and move the solar panels so they are facing the sun at all times.

The benefits of solar trackers are that they increase the amount of sunlight the solar panels can capture. They increase energy capture by up to 55% during the summer months and by around 15–20% during the winter months.

Unfortunately, the cost of these solar trackers means that they are rarely cost-effective. It is usually far cheaper to buy a larger solar array than it is to buy a solar tracker. Only if space is at a premium are solar trackers currently viable.

Solar array cables

Solar array cables connect your solar panels together and connect your solar array to the solar controller.

These cables are often referred to as 'array interconnects'. You can purchase them already made up to specified lengths or make them up yourself. The cables are extremely heavy duty and resistant to high temperatures and ultra-violet light. They also have a tough, extra-thick insulation to make them less prone to animal damage.

If you are planning to wire your solar array in parallel rather than in series, you need to ensure that your solar array cables can cope with the current that you are going to be generating through your solar array. If you are designing a parallel design system, I explain how you can calculate the size of cable required in the chapter on stand-alone system components. You can read about this on page 82.

Fuses and isolation switches

The ability to isolate parts of the system is important, especially while installing the system and carrying out maintenance. Even comparatively low voltages can be dangerous to work on.

Even small systems should incorporate a fuse between the batteries and the controller and/or inverter. If something goes wrong with the system, far better to blow a cheap fuse than fry a battery or a solar controller.

For all but the smallest systems, you will also need to incorporate isolation switches into your solar design. This will allow a battery bank to be disconnected for maintenance purposes. For any installation with more than one solar panel, and for all grid-connected systems, an isolation switch to disconnect the solar array should also be installed: I would recommend installing an isolation switch on the solar array for all solar arrays capable of generating over 100 watts of power.

If your solar panels are mounted some way from your inverter or controller, it can also be a good idea to have an isolation switch fitted next to the solar panels, as well as one fitted next to the inverter or controller. You can then easily disconnect the solar panels from the rest of the system for maintenance or in case of an emergency.

Ensure that the isolation switch you choose is capable of handling high-current DC circuits, with contacts that will not arc. Suitable isolation switches are available from any solar supplier.

If you are planning a grid-connected system, you will also need AC isolation switches to allow you to disconnect the inverter from the grid supply. You will require an isolation switch next to the inverter, and a second one next to the distribution panel.

Ground fault protection

Ground fault protection ensures that if there is a short within the solar array, the current flow is cut off immediately. This averts the risk of damage to either the controller or the solar array, and significantly reduces the risk of electrocution.

Ground fault protection works by measuring the current entering and exiting a circuit. If everything is working correctly, the current in should equal the current out. However, if there is a 'leak' or a partial short circuit, the system will see a difference in current and immediately shut down. A partial short circuit could occur if a solar panel was broken or if somebody touched an exposed cable.

Most solar inverters and solar controllers incorporate ground fault protection, using a *Residual Current Device* (RCD) built into the unit (note: RCDs are known as *Ground Fault Interrupters* – GFIs – in the United States and Canada). Many experts say that it is prudent to install a separate ground fault protector, even if the controller or inverter has ground fault protection built in. As the cost of an RCD or GFI is low and the benefits they provide are high, this is good advice.

You will require separate ground fault protection for your DC and AC circuits:

- For anything larger than 100-watt solar panel systems, and for all systems mounted to a building, you should install ground fault protection between your solar panels and your controller or inverter
- If you are installing a DC power supply into a building for running appliances, you must install ground fault protection between your controller and this power supply
- If you are using an inverter, you should install ground fault protection between your inverter and any load

There are specific RCD units for DC circuits and these are stocked by solar panel suppliers.

Components for Grid-Tie systems

Before discussing the components required for a grid-tie system in more detail, it is useful to look at how a grid-tie system is configured.

There are three basic designs for grid-tie solar energy systems:

- High-voltage in-series systems
- Low-voltage systems
- Micro-inverter systems

High voltage in-series

This is the most popular configuration for a grid-tie system today. Solar panels are connected together in series, producing a high-voltage DC power. This is then fed into a central inverter to convert the power into an AC source, which in turn is connected into the standard building electrical system:

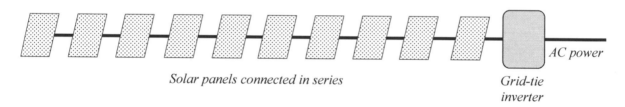

Solar panels connected in series   *Grid-tie inverter*

A simplified block diagram showing the basic layout of a high voltage in-series solar energy system

This design is the most cost-effective design for grid-tie systems. It is relatively straightforward to install, simple to maintain, and components are readily available. By running the solar array at high voltage, it is also very efficient, with minimal losses through the array itself and allowing the inverter to run at a very high level of efficiency. This is why this design is currently so popular within the grid-tie solar industry.

This high-voltage DC power has the benefit of great efficiency, but comes with a number of very significant safety risks. Voltages as high as 600 volts in North America and 1,000 volts in Europe are common: voltages that can very easily be fatal on contact.

These high voltages can also cause significant problems if there is damage to the wiring between solar panels, either due to a mistake during installation, through animal damage, or simply through wear over time. If a damaged cable generates a high-voltage, direct current electric arc whilst the panels are in direct sunlight, the immensely high temperatures can easily melt metal and are a potential fire hazard.

Connecting solar panels in series also has a significant disadvantage: when connected in series, the solar array is only as strong as its weakest link. If you have a damaged solar panel, shade blocking light to a few solar cells, or a damaged cable, the output of the entire solar array drops to the output of that weakest link.

Low voltage systems

It is for these reasons that, in previous editions of this book, I have advocated lower-voltage solar arrays for many grid-tie systems. This was achieved by running shorter series of solar panels and having multiple strings running in parallel. These are much safer than the very high voltage systems usually installed, lose little in efficiency and are inherently more tolerant of shade.

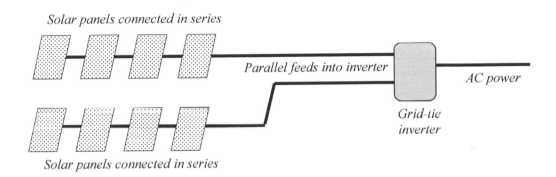

A simplified block diagram showing the basic layout of a low-voltage solar energy system where multiple strings of solar panels are connected in parallel

This design tends to be more expensive to install than an in-series design. You will either need to buy a grid-tie inverter that can accept multiple strings of solar panels, or buy a separate inverter for each string.

I no longer advocate this approach, because there is now another alternative available that, in my opinion, renders both high-voltage in-series and low-voltage systems redundant:

Micro-inverter systems

Micro-inverter systems have been around for a few years now, but until very recently had not gained popularity, mainly due to their higher cost.

However, with the huge growth in the popularity of grid-tie systems around the world, micro-inverter systems have become far more widespread as the benefits of this technology have become more apparent. Prices for micro-inverter systems are dropping fast and are now comparable to high-voltage in-series systems.

In a micro-inverter system, each solar panel has its own inverter, converting its output to AC power. In effect, each solar panel becomes its own independent solar energy system.

In the majority of micro-inverter systems, the inverter itself is mounted outside, bolted to the frame that also holds the solar panel in place. The individual solar panels are connected to an AC power cable that runs alongside the solar array and then feeds directly into the building's electrical system.

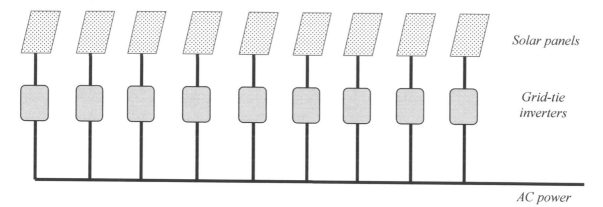

Solar panels

Grid-tie inverters

AC power

A simplified block diagram showing the basic layout of a micro-inverter system

There are some very significant benefits of micro-inverter systems over other forms of grid-tie systems:

- As each solar panel runs as an independent unit, if one panel is under-performing, either because of damage or simply because it is shaded, it does not affect the output from any of the other panels
- Because there is no high-voltage DC power running from one panel to the next, safety is less of an issue
- Installation, fault-finding and maintenance also become significantly easier
- It is easy to expand your grid-tie system in the future, as budget allows
- More flexible solar panel installation – you can have solar panels mounted in different locations and facing in different directions

Whilst today, series-connected solar systems with a central inverter are still the standard model for installing grid-tie systems, the industry is moving towards micro-inverter technology as a far better model.

Now you have a clearer idea of how a grid-tie system is put together, it is time to look at the components available in more detail.

Grid-tie solar panels

In the past, when most solar energy systems were stand-alone systems, almost every solar panel you could buy was rated for a 12-volt output. Whilst this is still true for smaller panels, there are now higher-voltage configurations available for larger solar panels.

As grid-tie systems have become more popular, higher-voltage solar panels have become available. Many solar panels of 150Wp capacity and over are rated for a 24-volt output and some manufacturers are now building solar panels with rated outputs of between 48 volts and 120 volts.

These higher voltages are well suited to grid-tie installations. By running your solar array at a higher voltage, you can keep the current flow low, which improves the efficiency of the overall system. Using high-voltage solar panels also gives you the option to connect multiple solar panels in parallel rather than in series, whilst retaining the benefit of the high-voltage current.

The 24-volt and 48-volt solar panels can work with many micro-inverter systems, too. Theoretically, this would allow a micro-inverter system to run more efficiently, although in practice the differences seem to be marginal.

Grid-tie inverters

Grid-tie inverters convert the DC power from your solar energy system into AC power, and convert the voltage to the same as the grid. This allows you to connect your system into the grid, enabling you to become a mini power station and supply your electricity to the electricity companies.

You cannot use an ordinary inverter for grid-tie applications. There are a number of reasons for this:

- Grid-tie inverters have to work in conjunction with the grid, in order to be able to export electricity to it. The AC pure sine waveform generated by the inverter has to be perfectly coordinated with the waveform from the grid
- There is an additional safety feature with grid-tie inverters to cut off power from the solar array if the grid shuts down
- Grid-tie inverters are connected directly to the solar panels. In an in-series system, this means the input voltage from the panels can fluctuate wildly, often jumping or dropping by several hundred volts in an instant. Non grid-tie inverters cannot cope with such massive voltage jumps
- In many countries, grid-tie inverters have to be certified for use with the grid

There are a number of things to consider when purchasing a grid-tie inverter:

- Input voltage
- Power rating
- Power tracking
- How many strings the inverter can support
- Diagnostics and reporting information
- Inbuilt safety systems
- Installation options and operating environment
- Certification and local regulations

Input voltage

Your choice of inverter will have a large voltage range in order to cope with the huge fluctuation of voltage that a solar array can provide. From this voltage range, you will be able to identify how many solar panels the inverter can cope with, when connected in series.

You need to remember that the rated voltage of a solar panel is not the maximum voltage that the solar panel can generate. The voltage from a single 12-volt solar panel can fluctuate anywhere from 12 volts on a very dull day, up to around 20 volts in intense overhead sunlight. If you have a 48-volt solar panel, or four 12-volt solar panels connected together in series, the voltage swing can be between 48 volts and 88 volts.

In addition to this, a solar panel can produce significantly higher voltages in an open circuit – i.e. when the solar array is generating power but the power is not being used. Depending on your solar panel, it is possible for a single 12-volt solar panel to generate 26 volts in an open circuit.

As you can see from the table below, the higher the nominal voltage from your solar array, the greater the voltage fluctuation can be:

Number of 12-volt solar panels	Nominal solar array voltage	Low voltage on dull day	Peak voltage in intense sunlight	Maximum open-circuit voltage
1	12-volt	12 volts	20 volts	26 volts
2	24-volt	24 volts	40 volts	52 volts
4	48-volt	48 volts	80 volts	104 volts
6	72-volt	72 volts	120 volts	156 volts
8	96-volt	96 volts	160 volts	208 volts
10	120-volt	120 volts	200 volts	260 volts
15	180-volt	180 volts	300 volts	390 volts
20	240-volt	240 volts	400 volts	520 volts
25	300-volt	300 volts	500 volts	650 volts
30	360-volt	360 volts	600 volts	780 volts
35	420-volts	420 volts	700 volts	910 volts
40	480-volts	480 volts	800 volts	1,040 volts

Note: the maximum open-circuit voltage allowable in the United States for grid-tied systems is 600 volts, whilst in Europe it is advisable that your system does not exceed 1,000 volts. You must ensure that your system never exceeds this.

As you can see from this table, if a heavy cloud blocks the sun on an otherwise clear day, you can see a voltage drop of several hundred volts in an instant. When the cloud passes over, the voltage shoots back up again.

It is important to ensure that the solar panel will work with the peak voltage of your solar array and not just the nominal voltage of your array. If you exceed the peak voltage of your inverter, the inverter will shut down to avoid damage. In extreme cases, you could damage or destroy your inverter by exceeding the input voltage rating.

In addition to the standard input voltage range, your inverter will also show a maximum voltage rating. This maximum voltage rating relates to the maximum *open circuit* voltage of your solar array. You must ensure that the open circuit voltage of your array does not exceed the maximum voltage of your inverter.

Power rating

There are two power ratings on a grid-tie inverter:

- Input power rating – the minimum and maximum amount of power the inverter can accept from the solar array
- Output power rating – the maximum amount of power and current the inverter can generate as an AC output

Input power rating

The input power rating shows the minimum and maximum wattage range the inverter can work with. In the main, the wider the range, the more efficient your inverter is.

The specification on an inverter will typically show three figures for input power rating:

- A nominal power rating, shown in watts
- A minimum and maximum power range
- A start-up power rating

The nominal power rating shows the maximum amount of power that the inverter can convert into an AC output. If you exceed this figure on your solar array, the additional power will be lost and converted into heat. Exceed this figure for long and your inverter may shut down to avoid overheating.

The minimum power rating shows the minimum amount of power that your solar array must generate in order for the inverter to start producing power. The maximum power rating shows the maximum amount of power that can be fed into the inverter before you risk damaging your inverter.

The start-up power rating is the minimum amount of power the solar inverter requires to power itself. If the solar array produces less than this amount of power, the inverter will switch off.

Because of the wide variation in the power a solar panel can produce, it is good practice to buy a bigger grid-tie inverter than you actually need. Remember that whilst a solar panel has a watt-peak (Wp) rating, in ideal conditions the panel itself may slightly exceed this rating in a real world environment.

Output power rating

The output power rating is the maximum continuous AC power that the inverter can generate. The output power information will show voltage, nominal output power in watts, the maximum output current in amps and the alternating current frequency.

For North America, the grid voltage is nominally set at 110 volts, with a frequency of 60 Hz. For the majority of the rest of the world, the grid voltage is nominally set at 230 volts, with a frequency of 50 Hz. In both cases, it is normal to get some variation in voltage and frequency.

The output power rating will also show the maximum efficiency rating of the inverter, given as a percentage. This rating is usually in the region of 90–94% with modern grid-tie inverters. If you are shopping on a budget, you must check this rating: some very cheap grid-tie inverters may be significantly less efficient.

Power tracking

As discussed on page 46, the efficiency of the solar array depends on how efficiently the fluctuating voltage is handled by your inverter.

If you are purchasing a grid-tie inverter, you should invest in one that incorporates maximum power point tracking (MPPT). Maximum power point tracking can provide an additional 15–20% of energy when compared to a non-MPPT inverter.

Today, MPPT is the norm, but there are a few older designs of inverter still on the market, often sold at bargain prices online. No matter how cheap these inverters are, the performance loss rarely makes them a worthwhile investment.

Multiple strings

A 'string' of solar panels is simply an array of solar panels connected in series. Some inverters allow you to connect more than one string of solar panels. These two strings then work as separate arrays, but feed the power through the same inverter.

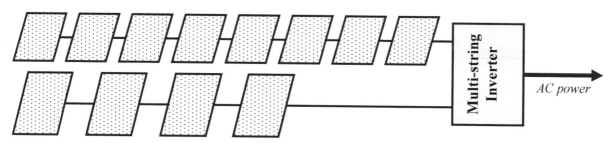

With multiple strings, you can mix and match solar panels and locate them in physically separate areas if you so wish. The two strings run completely independently of each other

With a multiple string system, your two solar arrays work independently from each other. This means that you can have two different sizes of array with different solar panels, or have the two arrays mounted at different orientations. If one array is partially shaded, the performance does not affect the second array.

This is different to connecting multiple strings together to create a serial/parallel hybrid. If you have multiple strings connected together, the two strings are still linked. You need to have identical setups on both strings and the panels need to be facing the same way. Failure to do this will result in lower solar performance.

Here we have two strings of solar panels, connected together in parallel to feed a single string inverter. The two strings have to match each other. If one string is compromised, for instance due to shading, the second string is also affected. (Note: you would not normally wire a grid-tie solar array in this way. Instead, you would connect everything in series on a single string.)

Multiple strings are of benefit in the following situations:

- Where you wish to fit two different sizes of solar panel
- Fitting solar panels facing at different orientations, such as on a roof that has two different pitches
- Resolving shading issues. Solar panels that are in shade at certain times of the day can be put onto a separate string so as not to affect the output of the rest of the system

Whilst an inverter that can handle multiple strings does have its benefits, there is usually an additional cost for multi-string inverters. If you are considering a multi-string inverter, you are probably better off choosing a micro-inverter system where every solar panel has its own inverter.

Diagnostics and reporting information

Almost all inverters provide a level of diagnostics and reporting information, either using a small display built into the inverter itself, a series of LEDs on the front panel of the inverter, a separate monitoring unit that plugs into the inverter, or by allowing you to connect a PC to the inverter.

Some inverters even have a built-in internet connection, allowing them to connect to a wireless network. This means your system can provide you with updates via e-mail, via a built-in website, or even send updates to your mobile phone. In some cases, these systems can be remotely monitored by the solar inverter supplier, who can then notify you if there are any potential issues with your system.

If you have a micro-inverter system with multiple inverters, the diagnostics and reporting system is usually a separate box, which is either connected to the AC power at the point where the solar array connects to your distribution panel, or it communicates wirelessly with the micro-inverters. This ensures that you have one central information point for your inverters.

As a bare minimum, you want an inverter that can tell you if your system is working and provide you with an indication of what may be the problem if a fault is detected. The diagnostics should be able to give you enough information to enable to you identify a fault with your system, such as:

- Insufficient or excess power from the solar array
- Grid connection issues
- Grid voltage or frequency issues
- Overheating

Most solar inverters will provide you with much more information, allowing you to see the voltage and current from your solar array and the amount of AC power being generated at that moment. They will also be able to show the amount of energy generated by the system, both for that day and since the system was installed.

Built-in safety

Most inverters incorporate safety shutdown systems as part of the inverter itself. It is common for inverters to have ground fault protection built in. As mentioned in the previous chapter, even if your inverter does provide this, it is still good practice to install additional ground fault protection when you design your system. This is incorporated into your system using a *Residual Current Device* (RCD). RCDs are known as *Ground Fault Interrupters* (GFIs) in the United States and Canada.

A grid-tie inverter will also monitor the power from the grid and shut down if it detects a power cut (sometimes referred to as 'Island Protection').

This power shutdown ensures that your solar energy system does not continue to feed power into the grid if there is a power cut. This is a necessary safety requirement: if workers are attempting to repair a power outage, they risk electrocution if power is being fed into the grid from your solar array while they are working.

Inverters should also shut down or derate if the internal temperature gets too high, in order to avoid permanent damage.

Installation options and operating environment

Inverters tend to be heavy units. They need to be mounted securely on a wall or bolted to the floor. They can generate a significant amount of heat, especially when running close to their rated output, and require good airflow around the units to keep them cool.

You can purchase inverters for either indoor or outdoor installation. If you are looking at installing an inverter outdoors, check that the inverter is sealed against dust and water ingress, rated to at least IP64.

Overheating inverters is the number one reason for grid-tie systems failing. As an inverter gets hotter, it provides less power, and if the temperature continues to rise, it will eventually shut down to avoid permanent damage. When choosing an inverter, check its operating temperatures and consider how you can ensure your system remains within these limits.

Inverters should always be installed in a well-ventilated area, away from the ceiling and with a clearance around each side, the top and the bottom. They cannot be installed in a sealed cupboard. Some inverters have the option of an external heat sink or temperature-controlled cooling fans to help keep the inverter cool.

Most inverters do make a small amount of noise. This is typically a continuous low-level hum. It is usually only noticeable if the surrounding area is quiet. However, for this reason, inverters are not usually installed inside the living space in a home or in an office environment. Instead, consider installing your inverter in a garage or on an outside wall of your building.

Occasionally, the sound made by the inverter has been known to resonate with the wall, amplifying the sound and making it quite unpleasant to live with. This can occur even if the inverter is mounted to an outside wall. This is a very rare occurrence, but is most likely to occur if you are planning to mount your inverter onto a wall made of solid concrete. The solution is to dampen the mounting between the inverter and the wall or floor that the inverter is mounted on. There is a very effective product called *Green Glue*, produced by the Green Glue Company (*www.GreenGlueCompany.com*) that is applied between the wall and the inverter. When it is compressed by remounting the inverter, the glue spreads out to form a sound isolation barrier that is particularly effective at blocking out low-resonance vibrations.

Buying from eBay

Some companies and individuals have been selling non-approved grid-tie inverters online, most commonly on eBay. These are often sold at a bargain price, bundled with a cheap solar panel and often advertised as a 'micro grid-tie system'.

The sellers claim that these systems are designed for amateur installation. The inverter plugs into the household electricity supply through a normal domestic power socket, and the systems look exceptionally easy to install and use. The sellers often claim that you can use these systems to sell power back to the utility companies and that they can be used to run the meter backwards.

These systems are highly dangerous and must be avoided. For a start, the equipment has inevitably not been certified for grid-tie use in any country. More importantly, the use of these systems is illegal in the United States, Canada, Australia and most of the European Community, because of the way the inverters connect to the household electricity supply, using a domestic power plug in reverse.

This means that the household plug has grid-level AC power running through it. This is extremely high risk and directly contravenes basic electrical safety legislation. In the UK, for instance, this is directly in contravention with BS7671:2008 (amd 1, 2011) 551.7.2 (ii). The catastrophic and potentially fatal results should somebody unplug the cable and accidentally touch the unshielded plug do not bear thinking about.

Never design any electrical system that risks grid-level AC power running through exposed connectors. The lives of the people around you are worth far more than saving a few pounds.

Components for Stand-Alone Systems

If you are designing a stand-alone system, there is a lot more design work and planning involved than there is for a similarly-sized grid-tie system. It is more critical to make sure your stand-alone system works: while a grid-tie system will not let you down if you do not generate enough energy, a stand-alone system will.

As well as considering and planning all the physical side of fitting the solar panels, routing the cabling and handling the safety aspects, you will also need to consider the voltage that your system will run at and design a battery system to store your energy.

Calculate your optimum voltage

Solar panels and batteries are normally both 12 volts, so logically you would think that it would make the most sense to run your system at 12 volts.

For small systems, you would be right. However, there are some limitations of 12-volt systems. Therefore, we now need to identify the optimum voltage for your system.

If you are still not comfortable with volts, watts, currents and resistance, now would be a good time to re-read Chapter 2: *A Brief Introduction to Electricity.*

Voltages and currents

Current is calculated as watts divided by volts. When you run at low voltages, your current is much higher than when you run at higher voltages.

Take a normal household low-energy light bulb as an example. A 12W light bulb running from grid-level voltages is consuming 12 watts of power per hour. The current required to power this light bulb at 230 volts is 0.05 amps (12W ÷ 230V = 0.05 amps) and at 110 volts is 0.1 amps (12W ÷ 110v = 0.1 amps).

If you run the same wattage light bulb from a 12-volt battery, you are still only consuming 12 watts of power per hour, but this time the current you require is 1 amp (12W ÷ 12V = 1 amp).

If you run the same wattage light bulb from a 24-volt battery, you halve the amps. You now only require ½ amp (12W ÷ 24V = ½ amp).

"So what?" I hear you say. *"Who cares? At the end of the day, we're using the same amount of energy, whatever the voltage."*

The issue is resistance. Resistance is the opposition to an electrical current in the material the current is running through. Think of it as friction on the movement of electrons through a wire. If resistance is too high, the result is power loss. By increasing your voltage, you can reduce your current and thereby reduce resistance.

You can counter the resistance by using thicker cabling, but you soon get to the point where the size of the cabling becomes impractical. At this point, it is time to change to a higher voltage.

What voltages can I run at?

For either a stand-alone or a grid fallback system, the most common voltages to run a solar electric system at are 12 volts, 24 volts or 48 volts.

As a rule, the most efficient way to run an electrical circuit is to keep your voltage high and your current low. That is why the grid runs at such high voltages: it is the only way to keep losses to a minimum over long distances.

However, you also need to factor cost into the equation: 12-volt and 24-volt systems are far cheaper to implement than higher voltage systems, as the components are more readily available, and at a lower cost. 12-volt and 24-volt devices and appliances are also easily available, whereas 48-volt devices and appliances are rarer.

It is unusual to go beyond 48 volts for stand-alone systems. Whilst you can go higher, inverters and controllers that work at other voltages tend to be extremely expensive and only suitable for specialist applications.

For a grid-tie system, you do have the option to run your solar array at a much higher voltage, by connecting lots of solar panels together in series. Grid-tie inverters are available that work anywhere from 12 volts up to 1,000 volts. In grid-tie systems, the voltage you run at depends on the number of solar panels you use.

How to work out what voltage you should be running at

Your choice of voltage is determined by the amount of current (amps) that you are generating with your solar array or by the amount of current (amps) that you are using in your load at any one time.

To cope with bigger currents, you need bigger cabling and a more powerful solar controller. You will also have greater resistance in longer runs of cabling, reducing the efficiency of your system, which in turn means you need to generate more power.

In our system, we are proposing a 12m (40 feet) long cable run from the solar array to the house, plus cabling within the house.

Higher currents can also reduce the lifespan of your batteries. This should be a consideration where the current drain or charge from a battery is likely to exceed $1/10^{th}$ of its amp-hour rating.

We will look at battery sizing later on, as current draw is a factor in choosing the right size of battery. It may be that you need to look at more than one voltage option at this stage, such as 12-volt and 24-volt, and decide which one is right for you later on. Finally, if you are planning to use an inverter to convert your battery voltage to a grid-level AC voltage, 12-volt inverters tend to have a lower power rating than 24-volt or 48-volt inverters. This can limit what you can achieve purely with 12 volts.

To solve these problems, you can increase the voltage of your system: double the voltage and you halve your current.

There are no hard and fast rules on what voltage to work on for what current, but typically, if the thickness of cable required to carry your current is over 6mm (and we'll calculate that in a minute), it is time to consider increasing the voltage.

How to calculate your current

As explained in Chapter 2, it is very straightforward to work out your current. Current (amps) equals power (watts) divided by volts:

$$\text{Power} \div \text{Volts} = \text{Current}$$
$$P \div V = I$$

Go back to your power analysis (our Example 1 is on page 33) and add up the amount of power (watts) your system will consume if you switch on every electrical item at the same time. In the case of our holiday home, if I had everything switched on at the same time, I would be consuming 169 watts of electricity.

Using the holiday home as an example, let us calculate the current based on both 12 volts and 24 volts, to give us a good idea of what the different currents look like.

Using the above formula, 169 watts divided by 12 volts equals 14.08 amps. 169 watts divided by 24 volts equals 7.04 amps.

Likewise, we need to look at the solar array and work out how many amps the array is providing to the system. We need a 320-watt solar array. 320 watts divided by 12 volts equals 26.67 amps. 320 watts divided by 24 volts equals 13.33 amps.

Calculating cable thicknesses

I will go into more detail on cabling later, but for now, we need to ascertain the thickness of cable we will need for our system.

For our holiday home, we need a 12m (40 feet) cable to run from the solar controller to the house itself. Inside the house, there will be different circuits for lighting and appliances, but the longest cable run inside the house is a further 10m (33 feet).

That means the longest cable run is 22m (72½ feet) long. You can work out the required cable size using the following calculation:

$$(L \times I \times 0.04) \div (V \div 20) = CT$$

L	Cable length in metres (one metre is 3.3 feet)
I	Current in amps
V	System voltage (e.g. 12V or 24V)
CT	Cross-sectional area of the cable in mm²

So calculating the cable thickness for a 12-volt system:

$$(22m \times 14.08A \times 0.04) \div (12V \div 20) = 20.65mm^2$$

Here is the same calculation for a 24-volt system:

$$(22m \times 7.04A \times 0.04) \div (24V \div 20) = 5.15mm^2$$

And just for sake of completeness, here is the same calculation for a 48-volt system:

$$(22m \times 3.52A \times 0.04) \div (48V \div 20) = 1.63mm^2$$

Converting wire sizes:

To convert cross-sectional area to American Wire Gauge or to work out the cable diameter in inches or millimetres, use the following table:

Cross-Sectional Area (mm²)	American Wire Gauge (AWG)	Diameter (inches)	Diameter (mm)
107.16	0000	0.46	11.68
84.97	000	0.4096	10.4
67.4	00	0.3648	9.27
53.46	0	0.3249	8.25
42.39	1	0.2893	7.35
33.61	2	0.2576	6.54
26.65	3	0.2294	5.83
21.14	4	0.2043	5.19
16.76	5	0.1819	4.62
13.29	6	0.162	4.11
10.55	7	0.1443	3.67
8.36	8	0.1285	3.26
6.63	9	0.1144	2.91
5.26	10	0.1019	2.59
4.17	11	0.0907	2.3
3.31	12	0.0808	2.05
2.63	13	0.072	1.83
2.08	14	0.0641	1.63
1.65	15	0.0571	1.45
1.31	16	0.0508	1.29
1.04	17	0.0453	1.15
0.82	18	0.0403	1.02
0.65	19	0.0359	0.91
0.52	20	0.032	0.81
0.41	21	0.0285	0.72
0.33	22	0.0254	0.65
0.26	23	0.0226	0.57
0.2	24	0.0201	0.51

| 0.16 | 25 | 0.0179 | 0.45 |
| 0.13 | 26 | 0.0159 | 0.4 |

From these figures you can see the answer straightaway. Our cable lengths are so great that we cannot practically run our system at 12 volts. The nearest match for 20.65mm² cables is 21.14mm². This is AWG 4 cable, with a cable diameter of 5.19mm. Cable this size is thick, heavy, inflexible, hard to source and very expensive.

This means we would need to lay extremely thick AWG 4 cables from the solar array and around our house to overcome the resistance. This would be expensive, inflexible and difficult to install.

Realistically, due to cable sizing, we are going to need to use either 24 volts or 48 volts for our solar electric system.

Mixing and matching solar panels

When specifying your solar array, you should keep to one type of panel rather than mixing and matching them. If you want a 100-watt array, for example, you could create this with one 100-watt solar panel, two 50-watt solar panels or five 20-watt solar panels.

If you do wish to use different solar panels in your array, you can do so by running two sets of panels in parallel with each other and either connecting them into a controller that can handle more than one feed, or by using more than one controller. This can be a useful way of creating the right wattage system, rather than spending more money buying bigger solar panels that generate more power than you actually need.

The result is slightly more complicated wiring, but it is often a more cost-effective solution to do this than to buy a larger capacity solar array than you actually need.

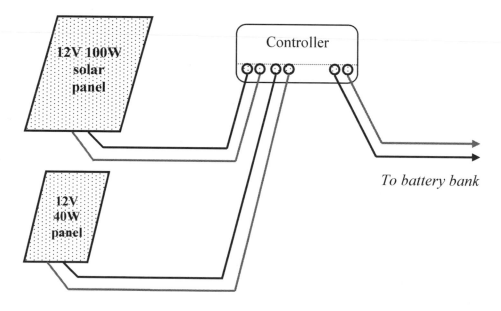

These two diagrams show different ways of connecting two solar panels of different sizes to the same system. Both these systems are running at 12 volts, using two different sized panels to create a 140 watt system

Previous page: This system uses a single controller. The controller has two separate input feeds. The two solar panels work independently of each other and the controller handles the mismatch in power output

Below: This system uses two controllers. The second controller is used to provide additional power to the batteries for the smaller solar panel.

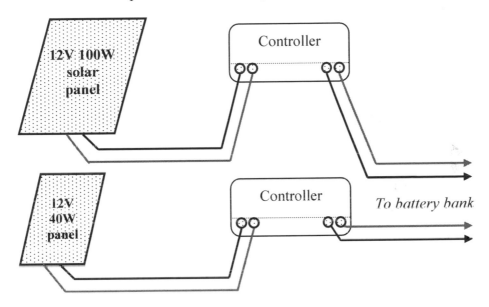

These solutions are effective if you are planning to start small and add to your solar energy system when needs and budget allows. It means that you can collect an assortment of solar panels over time and put them to good use within your one solar energy system.

If you end up with two different makes of solar panels with identical ratings, put them on their own separate circuits. Solar panels from different manufacturers are not all identical in their operating voltages or performances, so putting two different makes and models of solar panel on the same circuit is likely to compromise the performance of both panels, even if the specification of the two panels is similar.

If you buy multiple controllers for your solar energy system, to handle different makes, models and sizes of solar panel, you only need one main controller to handle the *power output* from the batteries. Your other controllers can be much cheaper and simpler pieces of equipment as they are only handling the power feed into the batteries. If you wish, you can even use a simple solar regulator that simply cuts off charging when the batteries are full, although these are usually not as efficient as a proper solar inverter.

Not all solar panels are 12-volt panels. Many solar panels are now designed predominantly for grid-tie installation only and are available in many different voltage configurations. Solar panels with voltage ratings of up to 120 volts are on the market, although the most common are 12-volt, 24-volt and 48-volt.

12-volt panels remain the most commonly available panels and are the most popular for stand-alone applications. If your stand-alone solar electric system runs at a voltage other than 12 volts, you can either install multiple solar panels in order to boost the system voltage, or choose higher-voltage solar panels. For example, if you wanted a 200-watt 24-volt solar array, you could achieve this in various ways, including:

- Using two 12-volt, 100-watt solar panels, connected in series
- Using one 24-volt, 200-watt solar panel

When choosing a solar array, you need to consider:

- The physical size: will it fit into the space available?
- The support structure: ready-made supports may only fit certain combinations of panels
- How much cabling you will need to assemble the array
- The system voltage: if you are not running at 12 volts, you will need multiple solar panels in order to build the system to the correct voltage as well as wattage

Batteries

There are a number of different options when it comes to batteries, and a number of specialist battery suppliers who can advise you on the best options for your solar installation.

Lead acid batteries usually come as either 6-volt or 12-volt batteries, although other voltages are also available. Batteries can be connected together in series to increase the voltage, or in parallel to keep the same voltage but increase the capacity.

The capacity of a battery is measured in amp-hours. The amp-hour rating shows how many hours the battery will take a specific drain: for instance, a 100-amp-hour battery has a theoretical capacity to power a 1-amp device for 100 hours, or a 100-amp device for 1 hour.

I say *theoretical*, because the reality is that lead acid batteries provide more energy when discharged slowly: a 100-amp-hour battery will often provide 20–25% less power if discharged over a five-hour period, compared to discharge over a twenty-hour period.

Secondly, a lead acid battery must not be run completely flat. A minimum of 20% state of charge (SOC) should be maintained in a lead acid battery at all times to ensure the battery is not damaged. For best overall battery life, you should design your system so that the battery charge rarely goes below 50%.

Types of batteries

There are three types of lead acid battery:

- 'Wet' batteries require checking and topping up with distilled water, but perform better and have a longer lifespan than other batteries

- AGM batteries require no maintenance but have a shorter overall life
- Gel batteries are also maintenance-free, do not emit hydrogen during charging and provide a reasonable overall life. They can be placed on their side or used on the move

In the past, most installers have recommended industrial quality 'wet' batteries for all solar installations. These provide the best long-term performance and the lowest cost. Often called *traction* batteries (as they are heavy-duty batteries used in electric vehicles), they can often have a lifespan of 8–10 years for a solar installation.

A lower cost option to the industrial-quality traction battery is the leisure battery, as used in caravans and boats. These are typically either wet batteries or AGM batteries. Their lifespan is considerably shorter than traction batteries, often requiring replacement after 3–4 years and significantly less in intensive applications.

The third option is the gel battery. These have the benefit of being entirely maintenance-free. They are also completely sealed and do not emit hydrogen gas. In the past, gel batteries have not been particularly reliable in solar installations, tending to require replacement after 1–2 years. However, more recently, smaller gel batteries have seen significant improvements in lifespan and they now are comparable to AGM batteries. The price has also dropped significantly.

Gel batteries are not suitable for big solar applications with a power drain of more than around 400 watt-hours, but they can provide an excellent, zero-maintenance alternative to wet batteries for smaller applications.

If your solar project requires batteries of 50 amp-hour capacity or less, gel batteries are a very good alternative to traction batteries.

Not all battery makes are the same. From my experience, the very best battery manufacturers for solar energy installations are Crown and Trojan, both of whom have excellent batteries specifically designed for solar installations. If only the very best will do and you are prepared to pay the premium, the Optima 'yellow top' batteries provide the benefits of AGM batteries in a smaller, lighter battery with some of the best overall performance figures of any battery available today.

Battery configurations

You can use one or more batteries for power storage. Like solar panels, you can wire your batteries in parallel in order to increase their capacity or in series in order to increase their voltage.

Unlike solar panels, which you can mix and match to create your array, you need to use the same specification and size of batteries to make up your battery bank. Mixing battery capacities and types will mean that some batteries will never get fully charged and some batteries will get discharged more than they should be. As a result, mixing battery capacities and types can significantly shorten the lifespan of the entire battery bank.

Battery lifespan

Batteries do not last forever, and at some stage in the life of your solar electric system, you will need to replace them. Obviously, we want to have a battery system that will last as long as possible and so we need to find out about the lifespan of the batteries we use.

There are two ways of measuring the lifespan of a battery, both of which tell you something different about the battery.

- Cycle Life is expressed as a number of cycles to a particular depth of discharge
- Life in Float Service shows how many years the battery will last if it is stored, charged up regularly, but never used

Cycle life

Every time you discharge and recharge a battery, you *cycle* that battery. After a number of cycles, the chemistry in the battery will start to break down and eventually the battery will need replacing.

The cycle life will show how many cycles the batteries will last before they need to be replaced. The life is shown to a 'depth of discharge' (DOD), and the manufacturers will normally provide a graph or a table showing cycle life verses the depth of discharge.

Typical figures that you will see for cycle life may look like this:

CYCLE LIFE	
20% DOD	1600 cycles
40% DOD	1200 cycles
50% DOD	1000 cycles
80% DOD	350 cycles

As you can see, the battery will last much longer if you keep your depth of discharge low.

For this reason, it can often be better to specify a larger battery, or bank of batteries, rather than a smaller set of batteries. Most experts recommend that you install enough batteries to ensure that your system does not usually discharge your batteries beyond 50% of their capacity.

The second benefit of a larger bank of batteries is that this gives you more flexibility with your power usage. If you need to use more electricity for a few days than you originally planned for, you know you can do this without running out of energy.

Holdover

When considering batteries, you need to consider how long you want your system to work while the solar array is not providing any charge at all. This time span is called *holdover*.

Unless you live inside the Arctic or Antarctic Circles (both of which provide excellent solar energy during their respective summers, incidentally), there is no such thing as a day without sun. Even in the depths of winter, you will receive some charge from your solar array.

You may find there are times when the solar array does not provide all the energy you require. It is therefore important to consider how many days holdover you want the batteries to be able to provide power for, should the solar array not be generating all the energy you need.

For most applications, a figure of between three days and five days is usually sufficient.

In our holiday home, we are deliberately not providing enough solar energy for the system to run 24/7 during the winter months. During the winter, we want the batteries to provide enough power to last a long weekend. The batteries will then be recharged when the holiday home is no longer occupied and the solar panel can gradually recharge the system.

For this purpose, I have erred on the side of caution and suggested a five-day holdover period for our system.

Calculating how long a set of batteries will last

Calculating how long a set of batteries will last for your application is not a precise science. It is impossible to predict the number of discharges, as this will depend on the conditions the batteries are kept in and how you use the system over a period of years.

Nevertheless, you can come up with a reasonably good prediction for how long the batteries should last. This calculation will allow you to identify the type and size of batteries you should be using.

First, write down your daily energy requirements. In the case of our holiday home, we are looking at a daily energy requirement of 695 watt-hours.

Then, consider the holdover. In this case, we want to provide five days of power. If we multiply 695 watt-hours a day by 5 days, we get a storage requirement of 3,475 watt-hours of energy.

Batteries are rated in amp-hours rather than watt-hours. To convert watt-hours to amp-hours, we divide the watt hour figure by the battery voltage.

If we are planning to run our system at 12 volts, we divide 3,475 by 12 to give us 290 amp-hours at 12 volts. If we are planning to run our system at 24 volts by wiring two batteries in series, we divide 3,475 by 24 to give us 145 amp-hours at 24 volts.

We do not want to completely discharge our batteries, as this will damage them. So we need to look at our cycle life to see how many cycles we want. We then use this to work out the capacity of the batteries we need.

On a daily basis during the spring, summer and autumn, we are expecting the solar array to recharge the batteries fully every single day: it is unlikely that the batteries will be discharged by more than 10–20%.

However, during the winter months, we could have a situation where the batteries get run down over a period of several days before the solar panels get a chance to top the batteries back up again.

So for four months of the year, we need to take the worst-case scenario where the batteries may get discharged down to 80% depth of discharge over a five day period and then recharged by the solar array.

The batteries will allow us to do this 350 times before they come to the end of their useful life.

350 cycles multiplied by 5 days = 1,750 days = 58 months

As this scenario will only happen during the four months from November to February, these batteries will last us for around 14½ years before reaching the end of their cycle life.

In reality, the *Life in Float Service* figure (i.e. the maximum shelf-life) for batteries is likely to be around ten years, which means that, for this application, they will fail before they reach their cycle life.

Based on our energy requirements of 145 amp-hours at 24 volts, and a maximum discharge of 80%, we can calculate that we need a battery capacity of 145 ÷ 0.8 = 181.25 amp-hours at 24 volts.

Second-hand batteries

There is a good supply of second-hand batteries available. These are often available as ex-UPS batteries (UPS = Uninterruptable Power Supplies) or ex-electric vehicle batteries.

Whilst these will not have the lifespan of new batteries, they can be extremely cheap to buy, often selling at their scrap value. If you are working to a tight budget and your power demands are not great, this is a very good way to save money.

Do not 'mix and match' different makes and models of batteries. Use the same make and model of battery throughout your battery bank. I would also advise against using a mixture of new and used batteries. This is a false economy as the life of your new batteries may be compromised by the older ones.

If you are considering second-hand batteries, try and find out how many cycles they have had and how deeply they have been discharged. Many UPS batteries have hardly been cycled and have rarely been discharged during their lives.

If buying ex-electric vehicle batteries, remember these have had a very hard life with heavy loads. However, ex-electric vehicle batteries can continue to provide good service for lower-demand applications: if your total load is less than 1kW, these batteries can provide good service.

If possible, try and test second-hand batteries before you buy them. Ensure they are fully charged up, and then use a battery load tester on them to see how they perform.

If your second-hand batteries have not been deep cycled many times, the chances are they will not have a very long charge life when you first get them. To 'wake them up', connect a solar controller or an inverter to them and put a low-power device onto the battery to drain it to around 20% state of charge. Then charge the battery up again using a trickle charge and repeat.

After three deep cycles, you will have recovered much of the capacity of your second-hand batteries.

If using second-hand batteries, expect them to provide half of their advertised capacity. So if they are advertised as 100-amp-hour batteries, assume they will only give you 50 amp-hours of use. In the case of ex-electric vehicle batteries, assume only one-third capacity.

The chances are, they will give you much more than this, but better to be happy with the performance of your second-hand batteries than to be disappointed because they are not as good as new ones.

Building your battery bank

Because we are running our system at 24 volts, we will need two 12-volt batteries connected in series to create our battery bank.

We therefore need two 12-volt batteries of 181.25 amp-hours each in order to create the desired battery bank.

It is unlikely that you are going to find a battery of exactly 181.25 amp-hours, so we need to find a battery that is *at least* 181.25 amp-hours in size.

When looking for batteries, you need to consider the weight of the batteries. A single 12-volt battery of that size will weigh in the region of 50kg (over 110 pounds)!

Safely moving a battery of that size is not easy. You do not want to injure yourself in the process. A better solution would be to buy multiple smaller batteries and connect them together to provide the required capacity.

Because batteries contain acid, they should be installed in a battery tray, so that any acid leaks may be contained. In Australia and Canada, regulations state that batteries must be enclosed in a ventilated, lockable and vermin-proof enclosure.

As it is not possible to buy 181.25 amp-hour batteries, I have decided to use four 100-amp-hour 12-volt batteries, giving me a battery bank with a total capacity of 200 amp-hours at 24 volts.

12-volt, 100-amp-hour batteries are still not lightweight. They can easily weigh 30kg (66 pounds) each, so do not be afraid to use more, lighter-weight batteries, if you are at all concerned.

To build this battery bank, you can use four 100-amp-hour, 12-volt batteries, with two sets of batteries connected in series, and then connect both series in parallel, as shown below:

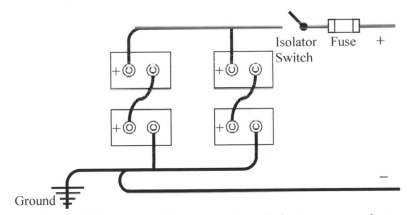

Four 100-amp-hour 12V batteries. I have paired up the batteries to make two sets of 100-amp-hour 24V batteries, and then connected each pair in parallel to provide a 200-amp-hour capacity at 24 volts.

If I were putting together a 12-volt battery system instead of a 24-volt battery system, I could wire together multiple 12-volt batteries in parallel in order to provide the higher capacity without increasing the voltage:

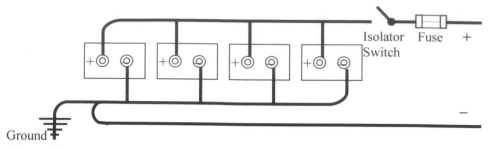

Four 100-amp-hour 12V batteries connected in parallel to provide a 400-amp-hour 12V battery bank.

Battery safety

When choosing batteries, you need to consider the safety aspects of batteries. With the exception of gel batteries, all lead acid batteries produce hydrogen, which needs to be ventilated. Batteries can also be very heavy and care is needed when lifting or moving them. Finally, due to the highly acidic nature of batteries, protective clothing should be worn whenever batteries are being worked on, and a chemical clean-up kit should be kept nearby.

I will go into more detail about handling batteries during the chapter on installation.

Solar controller

The solar controller looks after the batteries and stops them either being overcharged by the solar array or over-discharged by the devices running off the batteries.

Many solar controllers also include an LCD status screen so you can check the current battery charge and see how much power the solar array is generating.

Your choice of solar controller will depend on four things:

- System voltage
- The current of the solar array (measured in amps)
- The maximum current of the load (measured in amps)
- The level of detail you require from the status display

Some solar experts will sometimes add a fifth item to that list: battery type. To be fair, this was a problem with some older solar controllers, which only worked with specific battery types. Modern solar controllers work with all types of lead acid battery without a problem, although you may need to tell your solar controller what type of batteries you are using when you are setting up the system.

All but the very cheapest solar controllers provide basic information on an LCD screen that allows you to see how much power you have generated compared to how much energy you are using, and can also show the current charge stored in the battery. Some solar controllers include more detailed information that allows you to check on a daily basis how your power generation and usage compares.

Balancing the batteries

Another important function of a solar controller is to manage the charge in each battery and to ensure each battery is properly charged up.

As batteries get older, the charge of each battery will start to vary. This means that some batteries will charge and discharge at different rates to others. If left over time, the overall life of the batteries will deteriorate.

Intelligent solar controllers can manage these variations by balancing, or *equalizing*, the batteries they are charging. On most controllers, you need to manually activate a balance as part of a routine inspection.

Allow for expansion

When looking at solar controllers, it is worth buying one with a higher current rating than you actually need.

This allows you extra flexibility to add additional loads or additional panels to your solar array in the future without having the additional expense of replacing your solar controller.

Maximum power point tracking

More expensive solar controllers incorporate a technology called *maximum power point tracking* (MPPT). An MPPT controller adjusts the voltage being received from the solar array to provide the optimum voltage for charging the batteries without significant loss of watts from the voltage conversion.

If you have an MPPT controller, you can capture around 20% more of the power generated by the solar array compared to a more basic controller.

If you have less than 120W of solar panels, it can work out cheaper to buy extra solar panels rather than spend the extra money on an MPPT controller. However, prices continue to fall and, if you have the choice, a controller with maximum power point tracking is a worthwhile investment.

Ground fault protection

Many solar controllers include ground fault protection. In the case of a short from the solar array, a *Residual Current Device* (RCD) will cut off the current flow between the solar array and the controller, thereby averting the risk of damage to either the controller or the solar array.

In the United States and Canada, RCDs are also known as *Ground Fault Interrupters* (GFIs).

For anything larger than 100-watt solar panel systems, and for all systems mounted to a building, you need to incorporate a separate RCD/GFI into your system if you do not have ground fault protection built into your controller.

Backup power

Some controllers have one extra useful feature: the facility to start up an emergency generator if the batteries run too low and the solar array is not providing enough power to cope with the load.

This can be a useful facility for sites where the system must not fail at any time, or for coping with unexpected additional loads.

Whilst this may not seem so environmentally friendly, many generators are now available that run on bio-diesel or bio-ethanol. Alternatively, you can use an environmentally friendly fuel cell system instead of a generator. These tend to run on bio-methanol or zinc and only emit water and oxygen.

Using multiple controllers

Sometimes it is desirable to have multiple controllers on your solar energy system. For instance, you may want to install solar panels in different locations or facing in different directions, or you may have mismatched solar panels that you want to use (see page 86 for a block diagram). If you need to have multiple controllers, only one needs to have the expensive features such as battery balancing. The other controllers can be much simpler regulators that simply provide an additional charge to the batteries and switch off when the batteries are fully charged.

Inverters

We are not using an inverter with our holiday home, but many solar applications do require an inverter to switch up the voltage to grid-level AC current.

An inverter for stand-alone systems is a different piece of equipment to a grid-tie solar inverter. With a grid-tie inverter, your power is feeding into the grid and has to work in conjunction with the grid. The inverter connects directly to your solar panels and switches off when the solar panels no longer produce enough energy.

With a stand-alone system, your power is entirely separate from the grid. The inverter connects to your battery bank and switches off when the battery bank is running low on charge.

There are three things to consider when purchasing an inverter:

- Battery bank voltage
- Power rating
- Waveform

Battery bank voltage

Different inverters require a different input voltage. Smaller inverters, providing up to 3kW of power, are available for 12-volt systems. Larger inverters tend to require higher voltages.

Power rating

The power rating is the maximum continuous power that the inverter can supply to all the loads on the system. You can calculate this by adding up the wattages of all the devices that are switched on at any one time. It is worth adding a margin for error to this figure. Inverters will not run beyond their maximum continuous power rating for very long.

Most inverters have a peak power rating as well as a continuous power rating. This peak power rating allows for additional loads for very short periods of time, which is useful for some electrical equipment that uses an additional burst of power when first switched on (refrigeration equipment, for example).

As a general rule of thumb, go for a bigger power rating than you actually need. Inverters can get very hot when they get close to their maximum load for long periods of time. Many professionals recommend that you buy an inverter that has a continuous power rating that is at least one third higher than you plan to use.

Waveform

Waveform relates to the quality of the alternating current (AC) signal that an inverter provides.

Lower-cost inverters often provide a *modified sine wave* signal (sometimes advertised as a *quasi-sine wave*). More expensive inverters provide a *pure sine wave* signal.

Modified sine wave inverters tend to be considerably cheaper and also tend to have a higher peak power rating.

However, some equipment may not operate correctly with a modified sine wave inverter. Some power supplies, such as those used for laptop computers and portable televisions, may not work at all, while some music systems emit a buzz when run from a modified sine wave inverter.

These faults are eliminated with a pure sine wave inverter, which produces AC electricity with an identical waveform to the standard domestic electricity supply provided by the grid.

Installation options and operating environment

Small inverters with a continuous power rating of less than 3kW are lightweight units and are often simply placed on a shelf or a desk. Medium-sized inverters tend to be heavy units that need to be mounted securely on a wall. Larger inverters, rated at 10kW or above, may need to be bolted to a floor.

All inverters generate a significant amount of heat, especially when running close to their rated output, and require good airflow around the unit.

Most off-grid inverters are designed to be installed inside. Outdoor inverters are available, but they are expensive and may be difficult to source. If you are looking at installing an inverter outdoors, check that the inverter is sealed against dust and water ingress, rated to at least IP64.

Overheating inverters is the number one reason for any solar system failing. As an inverter gets hotter, they provide less power, and if the temperature continues to rise they will eventually shut down to avoid permanent damage. When choosing an inverter, check its operating temperatures and consider how you can ensure your system remains within these limits.

Inverters should always be installed in a well-ventilated area, away from the ceiling and with a clearance around each side, the top and the bottom. They cannot be installed in a sealed cupboard. Some inverters have the option of an external heat sink, or temperature-controlled cooling fans to help keep the inverter cool.

If your inverter is producing more than around 500 watts of power, it is likely to make a very small amount of noise. This is typically a continuous low-level hum. This is usually only noticeable if the surrounding area is quiet. However, for this reason, inverters are not usually installed inside the living space in a home or in an office environment. Instead, consider installing your inverter in a garage, or on an outside wall of your building.

Occasionally, with larger inverters, the sound made by the inverter has been known to resonate with the wall, amplifying the sound and making it quite unpleasant to live with, even if the inverter is mounted to an outside wall or the wall of a garage. This is a very rare occurrence, but is most likely to occur if you are planning to mount your inverter onto a wall made out of solid concrete. The solution is to dampen the mounting between the inverter and the wall or floor that the inverter is mounted onto. There is a very effective product called *Green Glue*, produced by the Green Glue Company (*www.GreenGlueCompany.com*) that is applied between the wall and the inverter. When it is compressed by remounting the inverter, the glue spreads out to form a sound isolation barrier that is particularly effective at blocking out low-resonance vibrations.

Ground fault protection

Most inverters now include ground fault protection. All inverters must always be grounded. If your chosen grid-tie inverter does not incorporate ground fault protection, you need to incorporate this into your system using a *Residual Current Device* (RCD). RCDs are known as *Ground Fault Interrupters* (GFIs) in the United States and Canada.

Cables

It is easy to overlook them, but cables have a vital part to play in ensuring a successful solar electric system.

There are three different sets of cables that you need to consider:

- Solar array cables
- Battery cables
- Appliance cabling

Solar array cabling has already been discussed (see page 70). For stand-alone systems, the battery and appliance cabling also needs to be correctly specified.

For all cabling, make sure that you always use cable that can cope with the maximum amount of current (amps) that you are planning to work with.

Take into account that you may wish to expand your system at some point in the future, and use a higher ampere cable than you actually need in order to make future expansion as simple as possible.

Battery cables

Battery cables are used to connect batteries to the solar controller and to the inverter. They are also used to connect multiple batteries together.

Battery interconnect cable is available ready-made up from battery suppliers, or you can make them up yourself. You should always ensure that you use the correct battery connectors to connect a cable to a battery.

Appliance cabling

If you are using an inverter to run your appliances at grid-level voltage, you can use standard domestic wiring, wired in the same way as you would wire them for connection to domestic AC power.

If you are running cabling for 12-volt or 24-volt operation, you can wire your devices up using the same wiring structure as you would use for grid-level voltage, although you may need to use larger cables throughout to cope with the higher current.

In a house, you would typically have a number of circuits for different electrical equipment: one for downstairs lighting, one for upstairs lighting and one or two for appliances, depending on how many you have. This has the benefit of keeping each individual cable run as short as possible, as well as reducing the amount of current that each circuit needs to handle.

As we have already learnt, low-voltage systems lose a significant amount of power through cabling. The reason for this is that the current (amps) is much higher and the power lost through the cable is proportional to the square of the current. You therefore need to keep your cable runs as short as possible, especially the cable runs with the highest current throughput.

I have already mentioned how you can calculate suitable cable thicknesses for your solar array earlier in this chapter. You use the same calculation for calculating cable thicknesses for appliance cabling.

Plugs and sockets

For 12-volt or 24-volt circuits with a current of less than 30 amps, you can use the same standard switches and light sockets as you do for normal domestic power.

However, you must not use the standard domestic plugs and sockets for attaching low-voltage devices to your low-voltage circuit. If you do, you run the risk that your low-voltage devices could accidentally be plugged into a high-voltage circuit, which could have disastrous consequences.

Instead, you have the choice of using non-standard plugs and sockets or of using the same 12-volt plugs and sockets as used in caravans and boats.

These low-voltage sockets do not need to have a separate earth (ground) wire, as the negative cable should always be earthed (grounded) on a DC circuit system.

Appliances

So far, I have talked a lot about 12-volt appliances, but you can buy most low-voltage appliances for either 12-volt or 24-volt and a lot of them are switchable between 12 and 24 volts.

Compared to appliances that run from grid-level voltages, you often pay more for low-voltage appliances. This is not always the case, however, and with careful shopping around, items like televisions, DVD players, radios and laptop computers need not cost any more to buy than standard versions.

Lighting

12-volt and 24-volt lighting is often chosen for off-grid solar electric systems, due to the lower power consumption of the lower-voltage lighting. You can buy low-voltage, energy-saving bulbs and strip lights, both of which provide the same quality of light as conventional lighting. Filament light bulbs are also available in low-voltage forms, and although these are not very energy-efficient, they do provide an excellent quality of light.

You can buy a lot of 12-volt lighting from ordinary hardware stores. Many kitchen and bathroom lights work at low voltage and will work just as well from a 12-volt battery supply as they will from the 12-volt AC transformers typically used with this lighting. Diachronic flood lamps, halogen spot lamps, strip lamps and LED lights often run at 12 volts, giving you an excellent choice. Buying these from a hardware store rather than from a specialist solar supplier can also save a considerable amount of money.

Refrigeration

A good selection of refrigerators and freezers are available that will run from 12-volt and 24-volt power supplies. Some refrigerators will run on both low-voltage DC and grid-level AC voltage, and some can run from a bottled gas supply as well.

Unlike most other devices that you will use, refrigerators need to run all the time. This means that, although the power consumption can be quite low, the overall energy consumption is comparatively high.

There are three types of low-voltage refrigerator available:

- *Absorption fridges* are commonly found in caravans and can often use 12-volt, grid-level voltage and bottled gas to power the fridge. These are very efficient when powered by gas, but efficiency when powered on lower voltages varies considerably for different models
- *Peltier effect coolers* are not really fridges in their own right; they are portable coolers, of the type often sold in car accessory shops and powered by the 12-volt in-car accessory socket. Whilst these are cheap, most of them are not very efficient. Avoid using these for solar applications
- *Compressor fridges* use the same technology as refrigerators in the home. They are the most efficient for low-voltage operation. They are more expensive than other types but their efficiency is significantly better: many models now consume less than 5 watts of electricity per hour

You can choose to use a standard domestic fridge for your solar electric system, running at grid-level voltages. However, they are typically not as efficient as a good 12-volt/24-volt compressor fridge. Domestic fridges also tend to have a very high starting current, which can cause problems with inverters.

A number of manufacturers now produce refrigerators that are specifically designed to work with solar power. Companies such as Waeco, Sundanzer and Shoreline produce a range of refrigerators and freezers suitable for home, medical and business use.

If you wish to use a standard domestic fridge, speak to the supplier of your inverter to make sure the inverter is suitable. Many refrigerators have a very high start-up current and you may need to buy a larger inverter that can handle this sudden demand.

Microwave ovens

Standard domestic microwave ovens consume a lot more power than their rated power: their rated power is output power, not input. You will find the input power on the power label on the back of the unit, or you will be able to measure it using a watt meter.

Typically, the input power for a microwave oven is 50% higher than its rated power.

Low-voltage microwave ovens are available, often sold for use in caravans and recreational vehicles (RVs). They tend to be slightly smaller than normal domestic microwaves and have a lower power rating, so cooking times will increase, but they are much more energy-efficient.

Televisions, DVDs, computer games consoles and music

Flat screen LCD televisions and DVD players designed for 12-volt or 24-volt operation are available from boating, camping and leisure shops. These tend to be quite expensive, often costing as much as 50% more than equivalent domestic televisions and DVD players.

However, many domestic LCD televisions (with screens up to 24-inch) and DVD players often have external power supplies and many of them are rated for a 12-volt input. Some investigations at your local electrical store will allow you to identify suitable models.

If you want to use one of these, it is worth buying a 12-volt *power regulator* to connect between the television and your battery. Battery voltages can vary between 11.6 volts and 13.6 volts, which is fine for most equipment designed for 12-volt electrics, but could damage more sensitive equipment. Power regulators fix the voltage at exactly 12 volts, ensuring that this equipment cannot be damaged by small fluctuations in voltage.

Many power regulators will also allow you to run 12-volt devices from a 24-volt circuit, and are much more efficient than more traditional transformers.

Power regulators also allow you to switch from one voltage to other low voltages, if required. For example, the Sony PlayStation 3 games console uses 8.5 volts, and with a suitable power regulator you can power one very effectively from 12-volt batteries.

Power regulators can step up voltages as well as step down. A suitable power regulator can switch the voltage from a solar battery bank to an output voltage of between 1½ volts and 40 volts, depending on the specification of the regulator.

This means that many normal household items with external power supplies, such as smaller televisions, laptop computers, DVD players, music systems and computer games, to name but a few, can be connected directly to your solar power system.

Music systems

Like televisions and DVD players, many music systems have an external power supply, and a power regulator can be used in place of the external power supply to power a music system.

Alternatively, you can build your own built-in music system using in-car components. This can be very effective, both in terms of sound quality and price, with the added benefit that you can hide the speakers in the ceiling.

Using a music system with an inverter which has a modified sine wave can be problematic. Music systems designed to run at grid-level voltages expect to work on a pure sine wave system and may buzz or hum if used with a modified sine wave inverter.

Dishwashers, washing machines and tumble dryers

Dishwashers, washing machines and tumble dryers tend to be very power hungry.

There are small washing machines, twin tubs and cool-air dryers available that run on low voltage, but these are really only suitable for small amounts of washing. They may be fine in a holiday home or in a small house for one person, but are not suitable for the weekly washing for a family of four.

If you need to run a washing machine from a solar electric system, you are going to need an inverter to run it. The amount of energy that washing machines consume really does vary from one model to the next. An energy-efficient model may only use 1,100 watts, whereas an older model may use almost three times this amount.

The same is true for dishwashers. Energy-efficient models may only use 500 watts, whereas older models may use nearer 2,500 watts. If you need to run a dishwasher, you will need to use an inverter.

Tumble dryers are hugely energy inefficient and should be avoided if at all possible. Most of them use between 2,000 and 3,000 watts of electricity and run for at least one hour per drying cycle.

There are various alternatives to tumble dryers. These range from the traditional clothes line or clothes airer to the more high-tech low-energy convection heating dryers that can dry your clothes in around half an hour with minimal amounts of power.

If you really must have a tumble dryer, you may wish to consider a bottled gas powered tumble dryer. These are more energy-efficient than electric tumble dryers and will not put such a strain on your solar electric system.

Air conditioning systems

Over the past couple of years, a number of manufacturers have been launching solar powered air conditioning and air cooling systems.

Air conditioning has traditionally been very power hungry. For this reason, solar powered air conditioning has been unaffordable, as a large solar array has been required simply to run the compressors.

In response, manufacturers have developed more efficient air conditioning systems, designed to run from a DC power source.

Companies such as Austin Solar, Solar AC, Securus, Sunsource, Sedna Aire, Hitachi and LG have all announced air conditioning units designed to work with solar energy.

Other manufacturers have developed evaporative air coolers that use a fraction of the power of an air conditioning unit. Whilst these air coolers do not provide the 'instant chill' factor of a full air conditioning system, by running constantly when the sun is shining, they can provide a very comfortable living and working environment at a fraction of the cost of full air conditioning.

Reputable brand names

Most solar manufacturers are not household names, and as such it is difficult for someone outside the industry to know which brands have the best reputation.

Of course, this is a subjective list and simply because a manufacturer does not appear on this list, it does not mean the brand or the product is not good.

Solar panel manufacturers and brands

Atlantis Energy, BP Solar, Canadian Solar, Clear Skies, EPV, Evergreen, Conergy, G.E. Electric, Hitachi, ICP, Kaneka, Kyocera, Mitsubishi, Power Up, REC Solar, Sanyo, Sharp, Solar World, Spectrolab, Suntech, Uni-solar.

Solar controller and inverter manufacturers and brands

Apollo Solar, Blue Sky, Enphase, Ever Solar, Exeltech, Fronius, Kaco, Magnum, Mastervolt, Morningstar, Outback, PowerFilm, PV Powered, SMA, Solectria, Sterling, Steca, SunnyBoy, Xantrex.

Battery manufacturers and brands

East Penn, Chloride, Crown, EnerSys, Exide, Giant, GreenPower, Hawker, ManBatt, Newmax, Odyssey, Optima, Panasonic, PowerKing, Tanya, Trojan, US Battery, Yuasa.

Shopping list for the holiday home

Because our solar electric system is being installed in the garden, 10 metres (33 feet) away from the house, we have worked out that we need to run our system at 24 volts rather than 12 volts, due to the high levels of losses in the system.

I have already calculated that I need 320 watts of power from my solar array at 24 volts. To achieve this, I will need to connect 12-volt solar panels in series to make a 24-volt system.

There are various different options available to make a 320-watt, 24-volt solar array. After checking with a number of suppliers, I have come up with the following options:

- Buy two 160-watt panels for a total of 320 watts of power – total cost £477 ($758 US)
- Buy four 80-watt panels for a total of 320 watts of power – total cost £339 ($540 US)
- Buy eight 40-watt panels for a total of 320 watts of power – total cost £400 ($636 US)
- Buy six 60-watt panels for a total of 360 watts of power) – total cost £468 ($744 US)

It is really worth shopping around and finding the best price. Prices can vary dramatically from one supplier to another and I have seen many cases where one supplier is selling a solar panel for over twice the price it is available from elsewhere.

Depending on what configuration I buy (and where I buy it), solar panel prices for the different combinations vary between £339 ($540 US) and £477 ($758).

Based on price and convenience, I have decided to go for the cheapest option and buy four Clear Skies 80-watt polycrystalline solar panels.

Because I am running at 24 volts and at a relatively low current, I have a good choice of solar controllers without spending a fortune. I decided to buy a Steca MPPT controller, which incorporates a built-in LCD display so I can see how much charge my batteries have at any one time. The cost of this controller is £200 ($318).

This drawing shows how I intend to wire up my solar array. I will pair two sets of panels together in series to bring the voltage up from 12 volts per panel to 24 volts per pair. I then connect the pairs together in parallel to maintain the 24 volts but to increase the power of the system to a total of 320 watts

I calculated that I needed 181Ah of 24-volt battery storage. I have decided to go for four Trojan 12V, 105Ah batteries, which I will connect together in pairs to provide me 210 Ah of power at 24 volts. The cost of these batteries is £496 ($788).

This drawing shows how I intend to wire up my batteries. I will pair two sets of batteries together in series to bring the voltage up to 24 volts per pair. I then connect the pairs together in parallel to maintain the 24 volts but to increase my storage capacity to 210 amp-hours at 24 volts

My Steca controller incorporates Ground Fault Protection, but I have decided to install a separate RCD (GFI) unit as well. It is a 'belt and braces' approach, but RCDs are extremely cheap and I feel it is worth the extra money. I still need a way of isolating the solar array manually. I choose to install three DC isolation switches: one between my controller and the solar array, one between my solar controller and my batteries and one between my controller and my distribution box. This allows me to isolate each part of my system separately, for maintenance or in case of an emergency.

For lighting, I have decided on 24-volt energy saving compact fluorescent light bulbs for inside use and a 24-volt halogen bulkhead light for an outdoor light. The energy saving compact fluorescent light bulbs look identical to grid-powered energy saving light bulbs and provide the same level of lighting as their grid-powered equivalents. Bulbs cost around £8/$13 each and I can use the same light switches and fittings as I would for lights powered by the grid.

I have decided to use a Shoreline RR14 battery-powered fridge, which can run on either 12-volt or 24-volt power supplies. This has a claimed average power consumption of 6 watts per hour and costs £380 ($610).

For television, I have chosen a Meos 19-inch flat screen TV with built-in DVD player. The Meos TV can run on 12-volt or 24-volt power and has an average power consumption of 45 watts. This is slightly higher than I was originally planning for (I was planning to buy a model with a 40 watt power consumption), but not by enough to be of any great concern.

At this stage, I now know the main components I am going to be using for my holiday home. I have not gone into all the details, such as cables and configuration. We need to complete that as we plan the detailed design for our solar energy system.

In conclusion

- When choosing solar panels, buy from a reputable manufacturer. The performance of the high-quality panels, especially in overcast conditions, is often better than the cheaper panels, and the improved build quality should ensure a longer life
- Lead acid batteries come in various types and sizes. You can calculate the optimum size of battery based on cycle life when operating on your system
- The voltage you run your system at will depend on the size of current you want to run through it. High-current systems are less efficient than low-current systems, and low-current inverters and controllers are inevitably cheaper
- Allow for future expansion in your system by buying a bigger controller and inverter than you currently need, unless you are absolutely certain your requirements are not going to change in the future
- Many appliances and devices are available in low-voltage versions as well as grid-level voltage versions. Generally, the low-voltage versions tend to be more efficient
- When wiring in 12-volt or 24-volt sockets, do not use standard domestic power sockets. If you do, you are running the risk of low-voltage devices being plugged into grid-level voltage sockets, which could have disastrous consequences

Planning, regulations and approvals

Depending on where you live around the world, there are different planning requirements, regulations and approvals needed for installing a solar energy system. Some countries have little or no regulation in place; other countries have extremely tight regulations. In some countries, the regulations change from one region to another.

Consequently, it is impossible to provide every bit of relevant information here. Instead, we provide much of this information on *www.SolarElectricityHandbook.com*. You will also be able to find information from your local planning authority and electricity providers.

Wherever you live around the world, there is a simple mantra for dealing with authority when it comes to building and electrical regulations and approvals: *if in doubt, ask*. Ignorance is never an excuse.

In the case of a solar installation, the people you need to speak to are your local planning office, your buildings insurance provider and, if you are building a grid-tie system, your local electricity company. Not only will they be able to help ensure you do not fall foul of any regulation; you will often find they are a helpful and useful source of information in their own right.

National and international standards for solar components

In the United States, Canada, Australia and across Europe, solar panels and inverters must comply with specific standards in order to be used in a grid-tie system. The units are tested to ensure that they conform to these standards before they are allowed on sale.

Across Europe, solar panels have to be certified to IEC safety standard IEC 61730 and performance standards IEC 61215 or IEC 61646. Solar grid-tie inverters have to conform to IEC 62109. Some European countries have additional certification. In Germany, grid-tie inverters must have a VDE126 certification, whilst in the United Kingdom, grid tie inverters that produce fewer than 16 amps of peak power (3.6kW) must have G83/1 certification, and larger inverters require the much more complicated G59/1 certification. Also in the United Kingdom, solar panels and inverters have to be certified by the Micro generation Certificate Scheme (MCS) in order to be eligible for feed-in tariffs and other financial incentives.

In the United States, solar panels, solar cables and inverters have to have UL certification. Solar panels must conform to the UL 1703 standard. Grid-tie inverters must conform to UL 1741 and solar cabling must conform to UL 4703 or UL 854 (USE-2). If you are using batteries in your design, the batteries must conform to either UL 1989, UL 2054, UL-SU 2580 or UL-SU 1973.

In Canada, solar panels must conform to safety standard ULC/ORD-C1703-1 and design standards CAN/CSA C61215-08 or CAN/CSA C61646-2, whilst grid-tie inverters must conform to CSA C22.2 No. 107.1. Batteries must conform to CAN/CSA F382-M89 (R2004).

In Australia, solar panels must conform to AS/NZS5033, whilst grid-tie inverters must conform to AS4777. If you are planning a stand-alone system in a building, your system must also conform to AS4509. If you are planning a mobile system, for instance in a caravan or recreational vehicle, your system must conform to AS3001.

It is worth noting that, in all of these regions, no differentiation is made between grid-tie solar and stand-alone systems for component selection. If you are building a solar energy system that is to be fitted to a building, your system must use certified components in order to comply with building and electrical safety regulations in these regions.

If you use non-approved equipment in a grid-tie system in these countries, you will not be allowed to connect your system to the grid. You are also likely to be in contravention of building regulations and may invalidate your buildings insurance.

Installation regulations

In many countries, including the United States, Canada, Australia, New Zealand and throughout most of the European Union, you cannot work on building electrics unless you are a qualified electrician. Some countries allow you to work on electrics, but your work has to be checked and certified by a qualified electrician before commissioning.

In the main, low-voltage DC circuits are excluded from this legislation, but it is worth ensuring that this is the case in your region.

In many countries, there are additional qualifications for electricians that allow them to install and certify solar energy systems. In most countries, it is not yet a legal requirement to have additional training in order to install photovoltaic systems.

However, if you wish to get access to government subsidies, feed-in tariffs or renewable energy certificates, you will almost certainly need to have your system installed, or at least checked, tested and certified, by qualified solar installation specialists. This is certainly the case in the United Kingdom and Australia. In the United States, subsidies vary from state to state, and often from county to county.

Getting your electricity supplier involved

If you are planning a grid-tie system, it is worth getting your electricity supplier involved in your project earlier rather than later. Sometimes they have their own requirements or lists of approved equipment. They often have specialists you can speak to, who can give you extra advice and support.

In most parts of the world, your electricity company will usually need to be involved while your system is being installed, replacing your current electricity meter with a specific import/export meter and carrying out the final inspection before approving your system.

Some electricity companies will only accept feed-in connections from professional solar PV installers. Almost all electricity providers insist that the installation is inspected and signed off, either by a certified solar installer or by one of their own inspectors, before they will accept your connection onto the grid.

Solar grants and selling your power

Around the world, governments are encouraging the take-up of solar energy. Financial assistance comes through various different schemes, and researching what is available can be confusing and time-consuming.

The different types of schemes that are offered in different places are described in more detail below. The specific schemes for grants and the amount of money you can receive for installing solar power and selling your electricity vary from country to country, and often from county to county. Many countries are currently reviewing their schemes, which means that information that is current one month will be out of date the next.

The Solar Electricity Handbook website has information on specific financial incentive schemes for various countries, as this can be kept more up-to-date than the book.

General information about grants, tax credits and feed-in tariffs

Whilst some schemes are flexible over who is allowed to install your solar energy system, most schemes work in conjunction with a governing body that insists that your system is installed by one of their members.

In some cases, individuals have been able to get their system signed off by a solar energy company in order to claim the financial incentives. However, this is often at the discretion of the solar installers, and many will refuse outright.

In general, financial incentives are being offered through four different mechanisms:

- Feed-in tariffs
- Tax credits
- Renewable Energy Certificates (RECs)
- Remote installation allowances

Feed-in tariffs

If you have a grid-tie solar energy system, you can often sell your power back to the electricity companies. This is done through a feed-in tariff, where the electricity providers agree to buy your surplus power at an agreed rate.

In some countries, feed-in tariffs are set by the electricity companies and can vary throughout the day, depending on supply and demand. In other countries, feed-in tariffs are fixed by the government, often at a premium rate in order to compensate solar owners for the up-front cost of installing their systems.

In many cases, the government guarantees the value of feed-in tariffs for a minimum number of years, thereby guaranteeing that owners make a return on their investments. A common theme with governments is to set a very high value for feed-in tariffs initially and then to reduce the feed-in tariff values for new customers significantly after two years. This has happened in Spain, Portugal, Germany, Hong Kong and now the United Kingdom. The message is this: if you are offered guaranteed long-term feed-in tariffs at a very good and guaranteed rate of return, take up the offer: the scheme is unlikely to remain available for more than two years.

Tax credits

A second option for compensating solar energy owners is a tax credit scheme where all or part of the installation cost of a solar energy system may be offset against tax. In some countries, these schemes are only available to businesses; in others they are available to individuals as well.

With a tax credit, you pay for the installation up front, but then receive part or all of the money back through tax credits over one, two or three years.

The United States of America are currently offering tax credits for people installing solar energy systems through the Federal Tax Credits for Consumer Energy Efficiency scheme. There are additional tax credits available in some states and counties.

Renewable Energy Certificates (RECs)

Renewable Energy Certificates are tradable certificates that prove that a certain amount of energy was generated from renewable sources. These certificates can be bought and sold on the open market. Whoever owns the certificate can claim to have bought electricity from a renewable resource.

There are two markets for buying renewable energy certificates:

- The voluntary sector – where individuals and companies elect to buy green electricity and pay a premium to do so
- The electricity providers themselves, who are mandated by governments to provide a certain percentage of their electricity from green sources

In some countries, individuals and small businesses who install solar energy systems are eligible to receive renewable energy certificates for the energy they produce. In many cases, renewable energy certificates are available for both grid-tie and stand-alone systems.

In some countries, governments have encouraged the take-up of small-scale solar by providing a multiplier for small solar generators. Under these schemes, solar energy owners can receive two, three or even five times the number of renewable energy certificates for the energy they produce, ensuring that owners can earn money from their small-scale solar energy systems.

Renewable energy certificates are also known as Green Tags in the United States and Tradable Renewable Certificates (TRCs) in South Africa and New Zealand.

Remote installation allowances

Remote installation allowances are offered in a few countries. They tend to be available for individuals and businesses with premises in remote areas, where the cost of connecting these buildings to the grid is very high.

By their very nature, these systems are stand-alone, off-grid applications. A good example of the sort of scheme on offer is the Australian model that provides credits for people installing solar where the cost of connecting their premises to the grid is greater than AUS $30,000, or the distance between the premises and the grid is greater than 1 km.

How much money is available, and how it is paid, varies from one scheme to another. In Australia, the remote installation allowance is offered by multiplying the number of renewable energy certificates that owners can receive for their system.

In conclusion

- There are different rules and regulations for installing solar power depending on where you live
- You have to comply with the building regulations and electrical regulations that are in force in your region
- You will be able to find help by talking to your local planning office and your electricity provider. You will also need to talk to your building insurance company
- There are many national and international standards for solar energy systems, covering both the actual physical hardware and how it is installed
- In many places around the world there are financial incentives available for solar energy providers. These schemes vary, but tend to fall into four camps: feed-in tariffs, tax credits, renewable energy certificates and remote installation allowances
- What incentives are available, and their value, changes regularly. Check the website for up-to-date information

Detailed Design

By now, you know what components you are going to use for your solar project. The next step is to work on your detailed design: effectively, a picture of what you want to build. Even for simple projects, it makes sense to draw up a diagram before installation.

The benefits of drawing a wiring diagram are numerous:

- It ensures that nothing has been overlooked
- It will assist in the cable sizing process
- It helps ensure nothing gets forgotten in the installation (especially where there is a group of people working together on site)
- It provides useful documentation for maintaining the system in the future

The wiring diagram will be different for each installation and will vary depending on what components are used. Read the product documentation for each component for information on how it must be wired.

If you have not yet chosen your exact components at this stage, draw a general diagram but make sure that you flesh this out into a detailed document before the installation goes ahead.

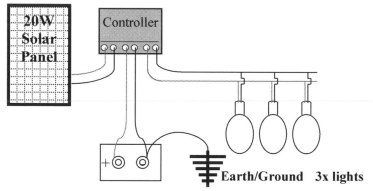

A sample wiring diagram for a simple stand-alone lighting system

When drawing up your wiring diagrams, you will need to remember the following:

Safety is designed in

It is easy to forget that solar energy can be dangerous. We are working with electricity and whilst any individual component may only be low-voltage, some of the currents involved can be quite significant. Furthermore, connecting multiple solar panels or batteries together in series can very quickly create a high voltage. It is therefore important that safety be taken into account during the detailed design phase of the project, as well as during installation.

When designing the system, ask yourself this question:

"What's the worst that can happen?"

Solar energy systems are relatively straightforward and the design of all the components you will use will keep risks to an absolute minimum. Nevertheless, there are potential risks. If you are aware of these risks, you can take steps to eradicate them in your design.

What is the worst that can happen with a solar installation?

With solar energy, we will be working in a few risk areas: DC electrics from the solar array, high currents from batteries, AC electrics if you are using an inverter, and high temperatures from the solar panels themselves.

Each of these risk areas can pose problems, both in isolation and when combined. It is worth considering these risks to ensure that you can design out as many of them as possible.

Grounding your electrics

Except for a very small system, such as rigging up a light in a shed, a solar energy system should always be earthed (grounded). This means running a wire from a negative terminal to an earthing rod (known as a *grounding rod* in North America) that is rammed into the ground.

An earthing rod (grounding rod) is a 1m (3 foot) long metal pole, typically made of copper. They are available from all electrical wholesalers and builders' merchants.

Connections to a ground prevent build-up of static electricity and can help prevent contact with high voltages if the circuit gets damaged.

If you are connecting a solar array to a home, you should always include a ground connection from the solar array itself. Whilst it is optional in other cases, it is always a good idea to include a ground from a solar array, where the array is capable of generating more than 200 watts. You must also earth the battery bank, as they are capable of delivering very high currents.

If you are using both AC electrics and DC electrics in your system, you must always have a separate ground for each system.

Grounding a system where you cannot connect to the ground

There may be instances where you are building a system where no connection to the ground is possible. For instance, a portable solar charging unit that can be carried anywhere, or a solar powered boat.

Typically, these designs are very small, using only DC electrics and running only a few amps of current. If your solar array is less than 100 watts, your system runs at 12 volts and you are drawing less than 10 amps of current, you are unlikely to need a common earth for all your components.

For larger systems, a *ground plane* is often used. A ground plane is a high-capacity cable connected to the negative pole on the battery, to which every other component requiring a ground is also connected. A thick, heavy-duty battery interconnection cable is often used as a ground plane cable, with thinner wires connecting to this ground plane cable from every other component requiring an earth.

As an alternative to a high-capacity cable, depending on what you are installing your solar system on, you can use a metal frame as the common ground for your system. In standard car electrics, for example, the ground plane is the car body itself.

DC Electrics

Direct current electricity typically runs at relatively low voltages: we are all familiar with AA batteries and low-voltage transformers used for charging up devices such as mobile phones. We know that if we touch the positive and negative nodes on an AA battery we are not going to electrocute ourselves.

However, direct current electricity can be extremely dangerous, even at comparatively low voltages. Around the world, a small number of people are killed every year by licking 9-volt batteries, because of the electric jolt they receive. Scale that up to an industrial grade heavy-duty 12-volt traction battery, capable of delivering over 1,000 amps of current, or a solar array capable of producing hundreds of volts on an open circuit, and it is easy to see that there is a real risk involved with DC electrics.

If you are electrocuted with AC power, the alternating current means that whilst the shock can be fatal, the most likely outcome is that you will be thrown back and let go. If you are electrocuted with DC power, there is a constant charge running through you. This means you cannot let go. If you are electrocuted with very high current DC, the injury is more likely to be fatal than a similar shock with AC power.

Because of the low current from a single solar panel, you are unlikely to notice any jolt if you short-circuit the panel and your fingers get in the way. However, wire up multiple solar panels together and it is a different story. Four solar panels connected in series produce a nominal 48 volts. The peak voltage is nearer 80–100 volts. At this level, a shock could prove fatal for a young child or an elderly person.

The current thinking with grid-tie solar systems is to connect many solar panels together in series, creating a very high-voltage DC circuit. Whilst there are some (small) efficiency benefits of running the system at very high voltage, there are risks as well, both during the installation and the ongoing maintenance of the system.

There are issues with the 12-volt batteries too. Industrial grade, heavy-duty batteries can easily deliver a charge of 1,000 amps for a short period. Short out a battery with a spanner and it will be red hot in just a few seconds. In fact, the current delivery is so great it is possible to weld metal using a single 12-volt battery.

The big risk with DC electrics is electrocuting yourself (or somebody else) or causing a short circuit, which in turn could cause a fire. Solar panels generate electricity all the time, often including a small current at night, and cannot simply be switched off. Therefore, there need to be manual DC circuit breakers (also called isolation switches) to isolate the solar panels from the rest of the circuit, plus a good ground and a ground fault protection system to automatically switch off the system should a short circuit occur.

If your system is running at a high voltage, you may want to consider multiple DC circuit breakers/ isolation switches between individual solar panels. This means that, as well as shutting off the overall circuit, you can reduce the voltage of the solar array down to that of a single panel or a small group of panels. This can be of benefit when maintaining the solar array, or in the case of an emergency.

A short circuit in a solar array can happen for many reasons. Sometimes it is because of a mistake during installation, but it can also occur as a result of general wear and tear (especially with installations where the tilt of the solar panels is adjusted regularly) or as a result of animal damage such as bird mess corroding cables or junction boxes, or a fox chewing through a cable.

Short circuits can also occur where you are using unsuitable cabling. Solar interconnection cabling is resistant to UV rays and high temperatures, and the shielding is usually reinforced to reduce the risk of animal damage. Always use solar interconnection cabling for wiring your array and for the cabling between the array and your solar controller or inverter.

When a short circuit does occur, there is often not a complete loss of power. Instead, power generation drops as resistance builds up. There is a build-up of heat at the point of failure. If you have a ground fault protection system such as an RCD or GFI in place, the system should switch itself off automatically at this point, before any further damage is caused.

If you do not have a ground fault protection system in place, the heat build-up can become quite intense, in some cases as high as several hundred degrees. There have been documented instances where this heat build-up has started a fire.

If a fire does break out, you need to be able to isolate the system as quickly as possible. Because a solar array cannot be switched off (it always generates power whenever there is light) there have been cases where the fire brigade have not been able to put out a fire generated by a fault in a solar array because there has been no way of switching it off. Isolating the solar array quickly, using a DC circuit breaker, resolves this problem.

However, remember that, even if you isolate the solar array, you are still generating power within the solar array. If you have many solar panels, the voltage and the current can still be quite considerable. The ability to shut down the array by fitting DC circuit breakers within the array can significantly reduce this power, rendering the system far safer if there is an emergency.

AC electrics

AC electrical safety is the same as household electrical safety. It is high-voltage and in many countries you are not allowed to work with it unless you are suitably qualified.

You will need to install two AC isolation switches: one switch between the inverter and the distribution panel to isolate the solar system completely, and one switch between your grid-feed and your distribution panel to isolate your system from the grid if you are running a grid-tie system.

If you are planning a grid-tie installation, you will need to speak with your electricity supplier, as there will often be additional requirements that you will need to incorporate. Your inverter will need to be a specific grid-tie system that switches off in the case of a grid power cut. This ensures that power is not fed back into the grid from your solar system in the case of a power failure, which could otherwise prove fatal for an engineer working on restoring power.

High temperatures

We have already touched on the risk of high temperatures with a solar array. Solar panels are black and face the sun: they can therefore get very hot on a warm day. It may not be hot enough to fry an egg, but in many climates it can certainly be hot enough to burn skin.

So make sure your solar array is installed in a place where it cannot be touched by curious children. If the solar panels are close to the ground, make sure there is some protection to keep people away from it.

The high temperatures become more of a problem if there is a fault within the solar array or with the wires running between solar panels. If a cable or a solar panel becomes damaged, there can be significant heat build-up. As already mentioned, this heat build-up can lead to a fire.

A *residual current device* (RCD), otherwise known as a *ground fault interrupter* (GFI) should avert this problem, allowing you to investigate the issue before significant damage occurs. However, manual DC circuit breakers should also be installed in order to override the system in case of an emergency.

Think safety

That is the end of the safety lecture for now. I will touch on safety again when we come to installation, but for now, please remember that safety does not happen by accident. Consider the safety aspects when you are designing your system and you will end up with a safe system. The additional cost of a few AC and DC circuit breakers, an earthing rod/ grounding rod, an RCD/ GFI and getting the right cables is not going to break the bank. It is money well worth spending.

Solar array design

All solar panels in an array must face in the same direction. This ensures that each cell receives the same amount of light, which is important for optimum power production.

Sometimes, you may want to install solar panels in different locations, such as on two different pitches of roof. In this instance, you need to keep the two banks of solar panels separate, running them as two separate arrays, either by feeding them into an inverter or controller that can handle more than one solar input, or by feeding them into two separate inverters or controllers.

If you wish to mix and match different sizes of solar panel, you will also need to set these up in separate arrays and wire these separately, either using an inverter or controller that can handle more than one solar input, or using two separate inverters or controllers.

If you are designing a grid-tie system, where you are considering different sizes or orientations of solar panels, you should seriously consider a micro-inverter system where each solar panel has its own inverter.

Solar array design – stand-alone systems

If you have more than one solar panel and you are running your solar electric system at 12 volts, then you will need to wire your panels together in parallel in order to increase your capacity without increasing the overall voltage.

If you are running your solar electric system at higher voltages, you either need to buy higher-voltage solar panels, or you will need more than one solar panel, wired in series to increase the voltage of the solar panels to the voltage of your overall system:

- For a 24-volt system, you have the choice of using 24-volt solar panels, or two 12-volt solar panels connected in series
- For a 48-volt system, you can use one 48-volt solar panel, two 24-volt solar panels connected in series, or four 12-volt solar panels connected in series

Once you have reached the voltage that you want, you can then run the panels both in series and in parallel, connecting strings of panels together in series to reach your desired voltage, and then connecting multiple strings together in parallel to increase your capacity:

IN SERIES
12V 100W panel + 12V 100W panel =
24V 200W array

IN PARALLEL
24V 200W array +
24V 200W array =
24V 400W array

A sample diagram of a 24-volt array where two sets of two 12-volt solar panels are connected in series in order to create a 24-volt array and the two arrays are then connected in parallel to create a more powerful 24-volt array.

Solar array design – grid-tie systems with micro-inverters

If you are designing a grid-tie system and using micro-inverters, your design is extremely simple. Each solar panel becomes a self-sufficient solar energy system, each feeding power into its own micro-inverter. The micro-inverters convert the energy to AC and feed it into the main AC circuit.

Solar panels

Grid-tie
inverters

AC power

Solar array design – grid-tie systems with a single inverter

If you are designing a grid-tie system with a single inverter, you will typically be connecting all your solar panels in series and then feeding this high-voltage DC power into an inverter.

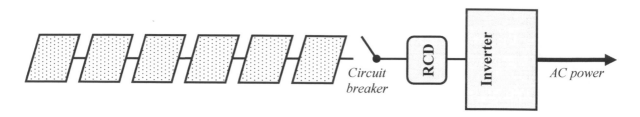

A simplified block diagram of a typical grid-tie system using a single inverter.
The residual current device (RCD) provides ground fault protection. In the
United States, an RCD is known as a ground fault interrupter (GFI)

Because of the very high DC voltages involved, additional safeguards are necessary. The solar array must always be grounded, there must be a DC circuit breaker (also known as an isolation switch) installed between the solar array and the inverter and there must be a DC residual current device/ ground fault interrupter installed to shut down the solar array in the case of a short circuit.

In the diagram on the following page, there are sixteen solar panels connected in series. Assuming each solar panel produces a 12-volt output, this system will run at a nominal 192 volts, with a peak power in the region of 320 volts and an open circuit voltage of 416 volts.

Because of the very high voltages, I have decided to install additional DC circuit breakers in the middle of the solar array in order to reduce the voltage within the array if I switch them off. This makes the system safer during maintenance and can reduce the risk of fire or electrocution in case of an emergency.

The diagram shows two AC isolation switches: one switch between the inverter and the distribution panel to isolate the solar system completely from your house and one switch to isolate your building from the grid.

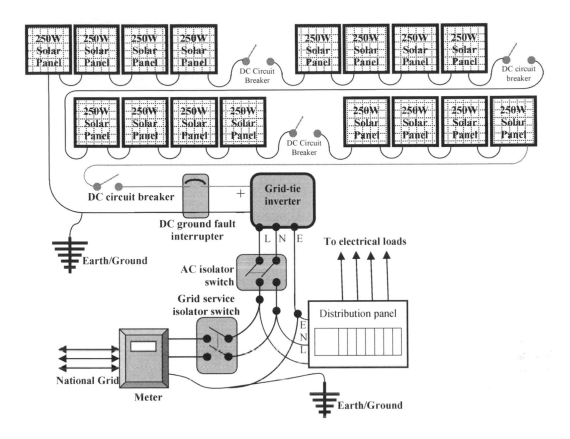

Above: A sample block diagram for a grid-tie system

In the United States, you are not allowed to have a grid-tie solar energy system where any part of that system has the potential to run at over 600 volts. This means that the open voltage of your solar array must be less than 600 volts. In general, this means that you will not want to connect more than twenty 12-volt solar panels or ten 24-volt solar panels in series, in order to ensure that you stay well below this level.

In Europe, it is advisable that your open circuit voltage remains under 1,000 volts. In general, this means that you will not want to connect more than thirty 12-volt solar panels, or fifteen 24-volt solar panels in series.

If you are running close to this limit, there are three options, which I have listed in order of preference:

- Install a micro-inverter system
- Install a multi-string system, either using an inverter that handles more than one solar feed, or by using two separate inverters
- Wire your solar panels in a parallel/series hybrid

119

For more information about open circuit voltages, refer back to page 76. To recap on multi-string systems, refer back to the section on Multiple strings on page 78.

Batteries

Batteries are wired in a similar way to your solar array. You can wire up multiple 12-volt batteries in parallel to build a 12-volt system with higher energy capacity, or you can wire multiple batteries in series to build a higher-voltage system.

When wiring batteries together in parallel, it is important to wire them up so that you take the positive connection off the first battery in the bank and the negative connection off the last battery in the bank.

This ensures equal energy drain and charging across the entire battery bank. If you use the same battery in the bank for negative and positive connections to the controller and inverter, you drain this first battery faster than the rest of the batteries in the bank. The first battery also gets the biggest recharge from the solar array.

This shortens the life of the battery and means all the batteries in the bank end up out of balance. Other batteries in the bank never get fully charged by the solar array, as the first battery will report being fully charged first and the controller will then switch the power off rather than continuing to charge the rest of the batteries in the bank. The result is that the batteries end up with a shorter lifespan.

*How to wire batteries in parallel: the diagram on the left, where power feed for both positive and negative is taken off the first battery in the bank, shows how **not** to do it – it will lead to poor battery performance and premature battery failure. The diagram on the right, where the positive feed is taken off the first battery in the bank and the negative feed is taken off the last battery in the bank, is **correct** and will lead to a more balanced system with a significantly longer life.*

Controller

A controller will have connections to the solar array, to the battery bank and to DC loads. Although controllers tend not to have the same heat problems as inverters, they can get warm in use. Make sure they are installed in an area with good ventilation around them and in a location where they can be easily checked.

Inverter

Where an inverter is used in a stand-alone or grid fallback system, it is connected directly to the battery bank and not through the controller.

Make sure that you design your system so that the inverter is in a well-ventilated area. Take into account the weight of the inverter and ensure that it is installed in a location where it can easily be checked.

Devices

Devices are connected to the inverter if they require grid-level voltage, or to the controller if they are low-voltage DC devices. They are never connected directly to the solar array or the batteries.

Specifics for a grid fallback system

Because a grid fallback system does not connect your solar energy system to the grid, you are less restricted as to the components you can use.

You must still adhere to basic wiring legislation for your country. In some countries (such as the United Kingdom, for instance) this can mean having the final connection into your building electricity supply installed by a fully qualified electrician, but this is significantly cheaper than having a grid-tie system installed and inspected.

The design for a grid fallback system is very similar to a stand-alone solar system, i.e. solar panels, solar controller and batteries. The only difference is what happens after the batteries.

The advantage of a grid fallback system is that it can work in three ways: it can provide power for an entire building, it can provide power for specific circuits within a building or it can provide power for a single circuit within a building.

More information and a sample circuit diagram for grid fallback configurations are included in Appendix E.

Circuit protection

Circuit protection is required in any system to ensure the system shuts down safely in the event of a short circuit. It is as valid on low-voltage systems as it is on high-voltage systems.

A low-voltage system can cause major problems simply because of the huge current that a 12-volt battery can generate: in excess of 1,000 amps in a short burst can easily cause a severe shock and even death or serious injury in some cases.

In the case of a short circuit, your wiring will get extremely hot and start melting within seconds unless suitable protection has been fitted. This can cause fire or burns, and necessary protection should be fitted to ensure that no damage to the system occurs as a result of an accidental short circuit.

Earthing (grounding)

In all systems, the negative terminal on the battery should be adequately earthed (referred to as *grounded* in North America). If there is no suitable earth available, a grounding rod or ground plane should be installed.

DC circuit protection

For very small systems generating less than 100 watts of power, the fuse built into the controller will normally be sufficient for basic circuit protection. In a larger system, where feed for some DC devices does not go through a controller, a fuse should be incorporated on the battery positive terminal.

Where you fit a fuse to the battery, you must ensure that all current from the battery has to pass through that terminal.

In DC systems with multiple circuits, it is advisable to fit fuses to each of these circuits. If you are using 12 volts or 24 volts, you can use the same fuses and circuit breakers as you would for normal domestic power circuits. For higher-voltage DC systems, you must use specialist DC fuses.

When connecting devices to your DC circuits, you do not need to include a separate earth (ground) for each device, as the negative is already earthed at the batteries.

Fit an isolation switch (DC disconnect switch) between your solar array and your inverter or controller. Fit a second isolation switch between your batteries and your controller and inverter.

Unless your controller or inverter already incorporates one, you should fit a DC residual current device/ ground fault interrupter between your solar array and your controller or inverter.

AC circuit protection

AC circuits should be fed through a distribution panel (otherwise known as a consumer unit). This distribution panel should be earthed (grounded) and should incorporate an earth leakage trip with a residual current device (RCD), otherwise known as a ground fault interrupter.

As you will have earthed your DC components, you must use a separate earth (ground) for AC circuits.

You must also install an AC disconnect switch (isolation switch) between your inverter and your distribution panel. In the case of a grid-tie system, this is normally a legal requirement, but it is good practice anyway.

The wiring in the building should follow normal wiring practices. You should use a qualified electrician for installing and signing off all grid-level voltage work.

Cable sizing and selection

Once you have your wiring diagram, it is worth making notes on cable lengths for each part of the diagram, and making notes on what cables you will use for each part of the installation.

Sizing your cables

This section is repeated from the previous chapter. I make no apologies for this, as cable sizing is one of the biggest mistakes that people make when installing a solar electric system.

Low-voltage systems lose a significant amount of power through cabling. This is because currents (amps) are higher to make up for the lack of voltage. Ohms law tells us that the power lost through the cable is proportional to the square of the current: the higher the current, the greater the resistance. To overcome this resistance, we must use thicker cables.

Wherever you are using low-voltage cabling (from the solar array to the controller, and to all low-voltage DC equipment) you need to ensure you are using the correct size of cable: if the cable size is too small, you will get a significant voltage drop that can cause your system to fail.

You can work out the required cable size using the following calculation:

$$(\text{Length} \times I \times 0.04) \div (V \div 20) = \text{Cable Thickness}$$

Length :	Cable length in metres (1m = 3.3 feet)
I :	Current in amps
V :	System voltage (e.g. 12 volts or 24 volts)
Cable Thickness:	Cross-sectional area of the cable in mm²

To convert this cross-sectional area to a cable diameter, or to an American Wire Gauge (AWG) size, refer to the table on page 85.

The cable thickness you are using should be at least the same size as the result of this calculation. Never use smaller cable, as you will see a greater voltage drop with a smaller cable, which could cause some of your devices not to work properly.

Protecting cable runs

When planning cable layouts, you need to ensure they are protected from unwanted attention from animals and children and from possible vandalism.

Rats and foxes chewing through cable insulation can be a big problem in some installations. This can be resolved by using rodent protected cabling. Using conduit is often a good idea as well, especially if you can use steel conduit or thin-wall electrical metallic tubing (EMT) to protect cables.

Designing your system to keep your cables runs as short as possible

If you have multiple devices running in different physical areas, you can have multiple cable runs running in parallel in order to keep the cable runs as short as possible, rather than extending the length of one cable to run across multiple areas.

By doing this, you achieve two things: you are reducing the overall length of each cable and you are splitting the load between more than one circuit. The benefit of doing this is that you can reduce the thickness of each cable required, which can make installation easier.

If you are doing this in a house, you can use a distribution panel (otherwise known as a consumer unit) for creating each circuit.

In the holiday home, for instance, it would make sense to run the upstairs lighting on a different circuit to the downstairs lighting. Likewise, it would make sense to run separate circuits for powering appliances upstairs and downstairs.

In the case of the holiday home, by increasing the number of circuits it becomes possible to use standard 2.5mm domestic 'twin and earth' cable for wiring the house, rather than more specialist cables. Not only does this simplify the installation, it keeps costs down.

Selecting solar cable

A common fault with poorly-designed or poorly-installed solar energy systems is under-performance where there is no clear source for the problem. In particular, this tends to occur around two to three years after the system was first installed.

The source of the problem is often either bird droppings or UV damage on cables leading from the solar panels to the inverters. This is usually caused by not using solar interconnection cables, which have a much tougher insulation that is UV protected, designed to withstand high temperatures and can withstand acidic bird droppings.

It is vital that you use specific solar interconnection cable to connect your solar panels together and for linking your solar panels to your inverter or controller. If you are not sure, look for cable that conforms to the UL 4703 or UL 854 (USE-2) specification for PV cabling. This is available from all solar equipment suppliers.

Controller cable

When calculating the thickness of cable to go between the controller and the battery, you need to take the current flow into the battery from the solar array as well as the flow out of it (peak flow into the battery is normally much higher than flow out).

Battery interconnection cables

You can buy battery interconnection cables with the correct battery terminal connectors from your battery supplier. Because the flow of current between batteries can be very significant indeed, I tend to use the thickest interconnection cables I can buy for connection between batteries.

Some sample wiring diagrams

As ever, a picture can be worth a thousand words, so here are some basic designs and diagrams to help give you a clearer understanding of how you connect a solar electric system together.

Above: A simple solar installation: a light with light switch, a small radio and a simple intruder alarm – perfect for a garden shed or a small lock-up garage. Because it is a small system, you may choose not to fit an isolation switch. Because the system is very small, I have decided only to fit a fuse between the controller and the battery

This is an interesting project – a solar powered river boat. Electric boats are gaining in popularity, thanks to their virtually silent running and lack of vibration. The only downside is recharging the batteries. Here, solar panels are used to recharge the batteries, charging them up during the week to provide all the power required for a weekend messing about on the river. The total cost of this complete system was less than the cost of a traditional outboard engine and fuel tank

Above: An example of a 12-volt solar system running an AC inverter to provide a normal building electricity supply in an off-grid installation. Below: the same system, wired at 24 volts

The holiday home wiring diagram

The next step

Once you have your wiring diagram, it is time to start adding cables, battery terminal clamps, fuses, isolation switches, earthing rods (referred to as *grounding rods* in North America) and, in this case, a distribution panel (otherwise known as a consumer unit) to your shopping list. It can help to add more detail to your wiring diagram as well, noting the locations of appliances and sockets, and the lengths of cables at each point.

Solar frame mounting

There are off-the-shelf solar array frames available, and your solar panel supplier will be able to advise you on your best solution.

Sometimes, however, these are not suitable for your project. In this case, you will either have to fabricate something yourself (angle iron is a useful material for this job) or get a bespoke mounting made specifically for you.

Solar panels in themselves are not heavy, but you do need to take into account the effect of wind loadings on your mounting structure. If the wind can blow underneath the solar array it will generate a 'lift', attempting to pull the array up off the framework. However, a gap beneath the solar array is useful to ensure the array itself does not get too hot. This is especially important in warm climates, where the efficiency of the solar panels themselves drops as they get hotter.

Making sure the mounting is strong enough is especially important as the solar array itself is normally mounted at an optimal angle to capture the noonday sun. This often means that, even if you are installing your solar array onto an existing roof, you may want to install the solar panels at a slightly different angle to the roof itself in order to get the best performance out of your system.

It is therefore imperative that your solar array mounting frame is strong enough to survive 20 years plus in a harsh environment and can be securely mounted.

If you are mounting your solar array on a roof, you must be certain that your roof is strong enough to take this. If you are not certain about this, ask a builder, structural surveyor or architect to assess your roof.

If you are planning to mount your solar array on a pole or on a ground-mounted frame, you will need to make plans for some good strong foundations. Hammering some tent pegs into the ground to hold a ground-mounted frame will not last five minutes in a strong wind, and a pole will quickly blow down if you only use a bucket of cement to hold it in place.

You should build a good foundation consisting of a strong concrete base on a compacted hardcore sub-base to hold a ground-mounted frame, and the frame itself should be anchored using suitable ground anchors, bolted using 25cm–30cm (10"–12") bolts.

For a pole, follow the advice given by the manufacturers. Typically, they need to be set in a concrete foundation that is at least 3 feet (1m) deep, and quite often significantly more.

To mount your solar panels onto your frame, make sure you use high-tensile bolts and self-locking nuts to prevent loosening due to wind vibration.

If your solar array is going to be easily accessible, you may wish to consider an adjustable solar mounting system so you can adjust the angle of tilt throughout the year. You can then increase the tilt during the winter in order to capture more winter sun, and decrease the tilt during the spring and summer in order to improve performance during those seasons.

For the holiday home project, the solar array is to be fitted to a specially constructed garden store with an angled roof. The benefit of this approach is that we can build the store at the optimum position to capture the sun. We can also install the batteries and solar controller very close to the solar array. In other words, we are creating an 'all in one' power station.

There are a number of regional shed and garden building manufacturers who will build a garden store like this to your specification. A good quality store, so long as it is treated every 2–3 years, will easily last 25–30 years.

If you go this route, make sure your chosen manufacturer knows what you are planning to use it for. You need to specify the following things:

- The angle of the roof has to be accurate in order to have the solar panels in their optimum position
- The roof itself has to be reinforced to be able to take the additional weight of the solar array
- The floor of the garden store (where the batteries are stored) must be made of wood. Batteries do not work well on a concrete base in winter
- There must be ventilation built into the store in order to allow the hydrogen gas generated by the batteries to disperse safely through the top of the roof
- The door to the garden store itself should be large enough for you to easily install, check and maintain the batteries
- You should consider insulating the floor, walls and ceiling in the garden store, either using polystyrene (Styrofoam) sheets or loft insulation. This will help keep the batteries from getting too cold in winter or too hot in summer

A garden store will still require a solid concrete foundation. Consideration of rainwater runoff is also important, to ensure the garden store does not end up standing in a pool of water.

Positioning batteries

You will have already identified a suitable location for your batteries. As discussed on the chapter on site surveys, your location needs to fit the following criteria:

- Water- and weather-proof
- Not affected by direct sunlight
- Insulated to protect against extremes of temperature
- Facilities to ventilate gases
- Protected from sources of ignition
- Away from children and pets

Lead acid batteries give off very small quantities of explosive hydrogen gas when charging. You must ensure that, wherever your batteries are stored, the area receives adequate external ventilation so that these gases cannot build up.

Because of the extremely high potential currents involved with lead acid batteries, the batteries must be in a secure area away from children and pets.

Do not install batteries directly onto a concrete floor. In extremely cold weather, concrete can cause an additional temperature drop inside the batteries that will adversely affect performance.

You need to ensure that your batteries are accessible for regular checks and maintenance. Many deep-cycle batteries require watering several times each year and connections must be checked regularly to ensure they have not corroded.

For all of the above reasons, batteries are often mounted on heavy-duty racking, which is then made secure using an open-mesh cage.

If you are installing your batteries in an area that can get very cold or very hot, you should also insulate your batteries. Extreme temperatures adversely affect the performance of batteries, so if your batteries are likely to be in an area where the temperature drops below 8°C (46°F) or rise above 40°C (104°F), you should consider providing insulation. If the temperature is likely to drop below freezing, you must provide it.

You can use polystyrene (Styrofoam) sheets underneath and around the sides of the batteries to keep them insulated. Alternatively, foil-backed bubble-wrap insulation (available from any DIY store in the insulation section) is even easier to use and has the benefit that it does not disintegrate if you ever get battery acid splashed on it.

Never insulate the top of the batteries, as this will stop them from venting properly and may cause shorts in the batteries if the insulating material you use is conductive.

Planning the installation

By now, you should have a complete shopping list for all the components you need. You should know where everything is to be positioned and what you need to in order to proceed.

Before placing any equipment orders, go back to your site and check everything one last time. Make sure that where you planned to site your array, controller, batteries and so on is still suitable and that you have not overlooked anything.

Once you are entirely satisfied that everything is right, place your orders for your equipment.

Bear in mind that some specialist equipment is often only built to order and may not be available straight away. If you require bespoke items such as solar mounting frames, or, as in the case of the holiday home, a complete garden store made up for mounting the solar panels and holding the batteries and controller, take into account that this could take a few weeks to be built for you.

In conclusion

- The detailed design ensures you have not overlooked any area of the design
- Consider the safety aspects of your system in your design. At each stage, ask yourself *"What is the worst that can happen?"* and then design around the problems
- The wiring diagram helps you envisage how the installation will work
- You need to keep cable runs as short as practically possible. You can do this by running several cables in parallel, either directly from the controller, through a junction box or through a distribution panel
- Splitting the cables into parallel circuits also means you reduce the current load on each circuit, thereby reducing resistance and improving the efficiency of your system
- If you are using an inverter to run at grid-level voltages, a qualified electrician is required to handle the electrical installation. However, your wiring diagram will help your electrician to envisage how your solar electric system should work
- You need to design your battery storage area to ensure your batteries can perform to the best of their ability

Installation

Congratulations on getting this far. If you are doing this for real, you will now have a garage or garden shed full of solar panels, batteries, cables, controllers, isolation switches, RCDs and whatnots. The planning stage is over and the fun is about to begin.

Before you get your screwdriver and drill out, there are just a few housekeeping items to get out of the way first…

Have you read the instructions?

No, of course you haven't. Who reads instructions anyway? Well, on this occasion, it is worth reading through the instructions that come with your new toys so that you know what you are playing with.

Pay particular attention to the solar controller and the inverter: there are many settings on most controllers and you need to make sure you get them right.

Safety

There are a few safety notices we ought to go through. Some of these may not be relevant to you, but read them all first, just to make sure.

Remember, you are working with electricity, dangerous chemicals and heavy but fragile objects. It is better to be safe than sorry.

Your First Aid kit

You will need a good First Aid kit on hand, including some items that you will not normally have in a regular First Aid kit. Most specifically, you will need an eye-wash and a wash kit or gel that can be applied to skin in case of contact with battery acid.

Chemical clean-up kit

You will be working with lead acid batteries that contain chemicals that are hazardous to health. You will require the following:

- A chemical clean-up kit suitable for cleaning up battery acids in the case of a spill
- A supply of strong polythene bags
- A good supply of rags/ disposable wipes to mop up any battery spillages

Chemical clean-up kits and chemical First Aid kits are available from most battery wholesalers and industrial tool suppliers. They only cost a few pounds. You probably will not need them but, if nothing else, they buy you peace of mind.

Considering the general public

If you are working in an area where the general public has access, you should use barriers or fencing, and signage to cordon off the area. Clear diversion signage should explain an alternative route.

In this scenario, I would recommend employing a professional team of builders to carry out the installation work on your behalf. They will already understand the implications of working in a public area and the relevant Health and Safety regulations.

Even if you do not have to consider the general public, you should still consider the people around you. Children love to get involved with these sorts of projects and there really can be some safety issues involved. Keep children out of the way, and let anyone in the vicinity know that you are working with high voltages and to keep away.

Working at height

You are very likely to be working at height and quite possibly crawling around on slanted rooftops.

Make sure you are using suitable climbing equipment (ladders, crawler boards, safety harnesses, scaffolding). You can hire anything that you do not have at reasonable prices.

If you have any concerns about working at heights, or if you are working beyond your area of competence at any time, remember there is no shame in hiring a professional. A professional builder can fit a solar array to a roof in 2–3 hours. This is typically less than half the time it takes an amateur DIY enthusiast.

Handling

Batteries, large inverters and solar arrays can be heavy. Solar panels themselves may not be heavy in their own right, but when several of them are mounted on a frame and then lifted they are heavy, bulky and fragile.

Moving and installing much of this equipment is a two-person job as a minimum. More people can be useful when lifting a solar array into position.

Working with batteries

Lead acid batteries are extremely heavy, in some cases weighing as much as an adult. Use proper lifting gear to move them, and look after your back.

Heavier batteries quite often have hoops in the top case. To lift a battery, I tend to use a piece of rope threaded through these hoops to create a carrying handle. This means I can carry a battery close to the ground and reduce the need to bend over to lift it.

Lead acid batteries contain acid. Unless they are gel batteries, the acid is in liquid form. It is extremely corrosive and extremely dangerous to health. Splashes of liquid from the batteries can cause severe chemical burns and must be dealt with immediately.

When working with lead acid batteries, stay safe:

- ALWAYS wear protective clothing, including overalls, eye protection (either protective glasses or a full-face shield) and protective gloves. I would also advise you to wear steel toe-capped shoes
- Keep batteries upright at all times
- Do not drop a battery. If you do, the likelihood is that the battery has been damaged. In the worst-case scenario, the casing could be cracked or broken
- If you drop a battery, place it immediately in a spill tray (a heavy-duty deep greenhouse watering tray can be used if necessary) and check for damage and leaks
- If you have a damaged battery, both the battery and the spill tray must be double-bagged in sealed polythene bags and marked as hazardous waste
- If you have a spillage from a battery, mop up the spillage immediately using rags or disposable wipes. Place these rags in a polythene bag, seal it and mark it as hazardous waste
- If any spillage from a battery comes into contact with clothing, remove clothing immediately and dispose of it in polythene bags
- If any spillage from a battery comes into contact with the eyes, wash repeatedly with eye-wash and seek urgent medical help
- If any spillage from a battery comes into contact with the skin, wash off immediately with water, apply an anti-acid wash, cream or gel to stop burning and then seek urgent medical help
- If you end up with battery acid in your mouth, wash your mouth out with milk. DO NOT swallow the milk. Spit it out. Then seek urgent medical help
- Do not smoke near batteries, and ensure that the area where you are storing the batteries is ventilated
- Prevent arcing or short circuits on battery terminals. Batteries can provide a huge current very quickly. Should you short-circuit a battery with a spanner, the spanner is likely to be red hot within a few seconds and could easily lead to fire or explosion. Remove any rings, bracelets or watches you may be wearing and keep tools a safe distance away from batteries

Gloves

You need two different sets of gloves for installing your solar array: a set of chemical gloves for moving batteries and a set of electrical protection gloves for wiring up your solar system.

When choosing suitable chemical gloves for working with batteries, consider the following:

- The gloves need to be quite strong, as lifting and moving batteries is hard on gloves
- A good grip is important

- Buy a glove with a medium or long cuff length, in order to protect both the hands and forearms
- The gloves should be made of a suitable material to protect against battery acid

The Health and Safety Executive website suggest that 0.4mm-thick neoprene gloves will give suitable protection through a full shift. If you do splash your gloves while working with batteries, make sure you wash them or replace them immediately, in order to avoid transferring acid to other parts of your body.

Electrically-insulated protection gloves give protection when working with high voltages. These are vitally important when working with high-voltage solar arrays and are recommended for all installations.

Electrically-insulated gloves come with different ratings to provide protection at different voltages:

- Class 00 gloves provide protection for up to 500 volts
- Class 0 gloves provide protection for up to 1,000 volts
- Class 1 gloves provide protection for up to 7,500 volts
- Class 2 gloves provide protection for up to 17,000 volts
- Class 3 gloves provide protection for up to 26,500 volts

For most solar installations, Class 00 or Class 0 gloves are the most appropriate. Remember that the open circuit voltage of a solar array can be more than double the nominal voltage of the solar array: twenty solar panels connected in series may only have a nominal voltage of 240 volts, but the open circuit voltage could be over 500 volts.

Like chemical gloves, choose gloves with a medium or long cuff length to protect both your hands and forearms.

If your electrically-insulated gloves are splashed with battery acid, remove and replace the gloves immediately.

All electrically-insulated gloves should be visually inspected and checked for tears and holes before use. Class 1–3 gloves require full electrical testing every six months.

Electrical safety

I make no apologies for repeating my mantra about electrical safety. Electrical safety is extremely important when installing a solar electric system.

Solar panels generate electricity whenever they are exposed to sunlight. The voltage of a solar panel on an 'open' circuit is significantly higher than the system voltage. A 12-volt solar panel can generate a 22–26 volt current when not connected.

Connect several solar panels in series and the voltage can get to dangerous levels very quickly: a 24-volt solar array can generate 45–55 volts, which can provide a nasty shock in the wrong circumstances, whilst a 48-volt solar array can easily generate voltages of 90–110 volts when not connected. These voltages can be lethal to anyone with a heart condition, or to children, the elderly or pets.

Solar systems produce DC voltage. Unlike AC voltage, if you are electrocuted from a direct current, you will not be able to let go.

Batteries can produce currents measuring thousands of amps. A short circuit will generate huge amounts of heat very quickly and could result in fire or explosion. Remove any rings, bracelets or watches you may be wearing and keep tools away from batteries.

The output from an inverter is AC grid-level voltage and can be lethal. Treat it with the same respect as you would any other grid-level electricity supply.

In many countries, it is law that if you are connecting an inverter into a household electrical system, you must use a qualified electrician to certify your installation.

Assembling your toolkit

As well as your trusty set of DIY tools, you will need an electrical multi-meter or volt meter in order to test your installation at different stages. You should use electrically-insulated screwdrivers whilst wiring up the solar array, and a test light circuit tester can be useful.

There are a few sundries that you ought to have as well:

- Cable ties are very useful for holding cables in place. They can keep cable runs tidy and are often good for temporary as well as permanent use
- A water- and dirt-repellent glass polish or wax, for cleaning solar panels
- Petroleum jelly is used on electrical connections on solar panels and batteries in order to seal them from moisture and to ensure a good connection

Preparing your site

As mentioned in the previous chapter, you may need to consider foundations for ground- or pole-mounting a solar array, or strengthening an existing roof structure if you are installing your solar array on a roof.

If you are installing your batteries in an area where there is no suitable earth (ground), you should install an earthing rod (grounding rod) as close to the batteries as is practical.

Testing your solar panels

Now the fun begins. Start by unpacking your solar PV panels and carry out a visual inspection to make sure they are not damaged in any way.

Chipped or cracked glass can significantly reduce the performance of the solar panels, so they should be replaced if there is any visible damage to the panel. Damage to the frame is not such a problem, so long as the damage will not allow water ingress to the panel and does not stop the solar panel from being securely mounted in position.

Next, check the voltage on the panel using your multi-meter, set to an appropriate DC voltage range.

Solar PV panels generate a much higher voltage on an 'open' circuit (i.e. when the panel is not connected to anything) than they do when connected to a 'closed' circuit. So do not be surprised if your multi-meter records an open voltage of 20–26 volts for a single panel.

Installing the solar array

Cleaning the panels

It is a good idea to clean the glass on the front of the panels first, using a water- and dirt-repellent glass polish or wax. These glass polishes ensure that rain and dirt do not stick to the glass, thereby reducing the performance of your solar array. They are available from any DIY store and many supermarkets and car accessories stores.

Assembly and connections

Some roof-mounted solar mounting kits are designed to be fitted to your roof before fitting the solar panels. Others are designed to have the solar panels mounted to the fixing kits before being mounted to the roof.

With a pole-mounted system, you typically erect your pole first and then fit the solar panels once the pole is in position.

A ground-based mounting system is the easiest to install, as there is no heavy lifting.

Typically, you mount and wire the solar panels at the same time. If you are stepping up the voltage of your system by wiring the panels in series, wire up the required number of panels in series first (i.e. sets of two panels for 24 volts, sets of four for 48 volts).

Once you have wired up a set of panels in series, test them using your multi-meter, set to a voltage setting to check that you have the expected voltage (20 volts plus for a 12-volt system, 40 volts plus for a 24-volt system and 80 volts plus for a 48-volt system).

Take care when taking these measurements, as 40 volts and above can give a nasty shock in the wrong circumstances.

Once you have wired each series correctly, make up the parallel connections and then test the entire array using your multi-meter, set to the appropriate voltage setting.

If you have panels of different capacities, treat the different sets of panels as separate arrays. Do not wire panels of different capacities together, either in series or parallel. Instead, connect the arrays together at the controller.

Once you have completed testing, make the array safe so that no one can get an electric shock by accident from the system. To do this, connect the positive and negative cables from the solar array together to short-circuit the array. This will not damage the array and could prevent a nasty shock.

Roof-mounting a solar array

If you are roof-mounting a solar array, you will normally have to fit a rail or mounting to the roof before attaching the solar array.

Once this is in place, it is time to fit the array itself. Make sure you have enough people on hand to be able to lift the array onto the roof without twisting or bending it. Personally, I would always leave this job to professional builders, but the best way seems to be to have two ladders and two people lifting the array up between them, one on each ladder, or using scaffolding.

Final wiring

Once your solar array is in position, route the cable down to where the solar controller is to be installed. For safety purposes, ensure that the cables to the solar array remain shorted whilst you do this.

If you are installing a DC isolation switch and a residual current device (known as a ground fault interrupter in North America), install them between the solar array and the controller.

Once you have the cables in position, un-short the positive and negative cables and check with a meter to ensure you have the expected voltage readings. Then short the cables again until you are ready to install the solar controller.

Installing the batteries

Pre-installation

Before installing the batteries, you may need to give them a refresher charge before using them for the first time.

You can do this in one of two ways. You can use a battery charger to charge up the batteries, or you can install the system and then leave the solar panels to fully charge up the batteries for a day or so before commissioning the rest of the system.

Put a sticker on each battery with an installation date. This will be useful in years to come for maintenance and troubleshooting.

Positioning the batteries

The batteries need to be positioned so they are upright, cannot fall over and are away from members of the public, children and any sources of ignition.

For insulation and heating purposes, batteries should not be stood directly on a concrete floor: during the winter months, a slab of concrete can get extremely cold and its cooling effects can have detrimental effects on batteries. I prefer to mount batteries on a wooden floor or shelf.

Ventilation

If there is little or no ventilation in the area where the batteries are situated, this must be implemented before the batteries are sited.

As batteries vent hydrogen, which is lighter than air, the gas will rise up. The ventilation should be designed so that the hydrogen is vented out of the battery area as it rises.

Access

It is important that the battery area is easily accessible, not just for installing the batteries (remembering that the batteries themselves are heavy), but also for routinely checking the batteries.

Insulation

As mentioned earlier, if you are installing your batteries in an area that can get very cold or very hot, you should insulate your batteries.

Polystyrene (Styrofoam) sheets or foil-backed bubble-wrap can be used underneath and around the sides of the batteries to keep them insulated. DO NOT INSULATE THE TOP OF THE BATTERIES as this will stop them from venting properly and may cause shorts in the batteries if the insulating material you use is conductive.

Connections

Once the batteries are in place, wire up the interconnection leads between the batteries to form a complete battery bank.

Always use the correct terminals for the batteries you are using and make sure the cables provide a good connection. You should use battery interconnection cables professionally manufactured for the batteries you are using.

Use petroleum jelly around the mountings to seal it from moisture and ensure a good connection.

Next, add an earth (ground) to the negative terminal. If there is no earth already available, install an earthing rod (grounding rod) as close as possible to the batteries.

Now check the outputs at either end of the batteries using a multi-meter to ensure you are getting the correct voltage. A fully-charged battery should be showing a charge of around 13–14 volts per battery.

Installing the control equipment

The next step is to install the solar controller and the power inverter if you are using one.

Mount these close to the batteries. Ideally they should be mounted within a metre (3 feet), in order to keep cable runs as short as possible.

Most solar controllers include a small LCD display and a number of buttons to configure the controller. Make sure the solar controller is easily accessible and that you can read the display.

Some solar controllers that work at multiple voltages have a switch to set the voltage you are working at. Others are auto-sensing. Either way, check your documentation to make sure you install the solar controller in accordance with the manufacturer's instructions. If you have to set the voltage manually, make sure you do this now, rather than when you have wired up your system.

Inverters can get very hot in use and adequate ventilation should be provided. They are normally mounted vertically on a wall, in order to provide natural ventilation. The installation guide that comes with your particular make of inverter will tell you what is required.

Some inverters require an earth (ground) in addition to the earth on the negative terminal on the battery. If this is the case, connect a 2.5mm² green-yellow earth cable from the inverter to your earth rod (ground rod).

If you are installing a DC isolation switch between the solar panels and your control equipment, connect that up first and make sure it is switched off.

Once you have mounted the controller and inverter, connect the negative cables to the battery, taking care that you are connecting the cables to the correct polarity. Then un-short the positive and negative cables from the solar array and connect the negative cable from the solar array to the solar controller, again taking care to ensure the cable is connected to the correct polarity.

Now double-check the wiring. Make sure you have connected the cables to the right places. Double-check that you have connected your negative cable from your solar array to the negative solar input connection on your solar controller. Then double-check that you have connected your negative cable from your battery to the negative battery input on your solar controller and your inverter. Only then should you start wiring up your positive connections.

Start with the battery connection. If you are planning to install a fuse and DC isolation switch into this cable, make sure that your fuse and switch work for both the solar controller and the inverter (if you are using one). Connect the inverter and the solar controller ends first and double-check that you have got your wiring correct, both visually and by checking voltages with a volt meter, before you connect up the battery bank.

Finally, connect up the positive connection from your solar array to the solar controller. At this point, your solar controller should power up and you should start reading charging information from the screen.

Congratulations. You have a working solar power station!

Installing a grid-tie system

Before starting to install your grid-tie system, you must have already made arrangements with your electricity provider for them to set you up as a renewable energy generator.

Regulations and agreements vary from region to region and from electricity provider to electricity provider, but at the very least they will need to install an export meter to your building in order to accurately meter how much energy you are providing. They will also ask for an inspection certificate from a qualified electrician, to confirm that the work has been done to an acceptable standard.

Physically, installing a grid-tie system is very similar to installing any other solar energy system, except, of course, you do not have any batteries to work with.

However, you do have to be careful while wiring up the high-voltage solar array. When the solar array is being connected up you can have a voltage build-up of several hundred volts, which can quite easily prove fatal. If building a high-voltage array, cover the solar panels while you are working on them and wear electrically-insulated gloves at all times.

Commissioning the system

Once you have stopped dancing around the garden in excitement, it is time to test what you have done so far and configure your solar controller.

Programming your solar controller

The type of solar controller you have will determine exactly what you need to configure. It may be that you do not need to configure anything at all, but either way you should check the documentation that came with the solar controller to see what you need to do.

Typically, you will need to tell your solar controller what type of batteries you are using. You may also need to tell your solar controller the maximum and minimum voltage levels to show when the batteries are fully charged or fully discharged. You should have this information from your battery supplier, or you can normally download full battery specification sheets from the internet.

Testing your system

You can test your solar controller by checking the positive and negative terminals on the output connectors on the controller using your multi-meter. Switch the multi-meter to DC voltage and ensure you are getting the correct voltage out of the solar controller.

If you have an inverter, plug a simple device such as a table lamp into the AC socket and check that it works.

If your inverter does not work, switch it off and check your connections to the battery. If they are all in order, check again with a different device.

Charging up your batteries

If you have not carried out a refresher charge on your batteries before installing them, switch off your inverter and leave your system for at least 24 hours in order to give the batteries a good charge.

Connecting your devices

Once you have your solar power station up and running and your batteries are fully charged, it is time to connect your devices.

If you are wiring a house using low-voltage equipment, it is worth following the same guidelines as you would for installing grid-voltage circuits.

For low-voltage applications, you do not need to have your installation tested by a qualified electrician, but many people choose to do so in order to make sure there are no mishaps.

The biggest difference between AC wiring and DC wiring is that you do not need to have a separate earth (ground) with DC electrics, as the negative connection is earthed both at the battery and, if you are using one, at the distribution panel.

If you are using 12-volt or 24-volt low-voltage circuits, you can use the same distribution panels, switches and light fittings as you would in a grid-powered home. As already suggested, do not use the same power sockets for low-voltage appliances as you use for grid-powered appliances. If you do, you run the risk that low-voltage appliances could be plugged directly into a high-voltage socket, with disastrous consequences.

In conclusion

- Once you have done all your preparation, the installation should be straightforward
- Heed the safety warnings and make sure you are prepared with the correct safety clothing and access to chemical clean-up and suitable First Aid in case of acid spills
- Solar arrays are both fragile and expensive. Look after them
- The most likely thing that can go wrong is wiring up something wrongly. Double-check each connection
- Check each stage by measuring the voltage with a multi-meter to make sure you are getting the voltage you expect. If you are not, inspect the wiring and check each connection in turn

Troubleshooting

Once your solar electric system is in place, it should give you many years of untroubled service. If it does not, you will need to troubleshoot the system to find out what is going wrong and why.

Keep safe

All the safety warnings that go with installation also relate to troubleshooting. Remember that solar arrays will generate electricity almost all the time (except in complete darkness), and batteries do not have an 'off' switch.

Common faults

Unless you keep an eye on your solar energy system, problems can often go undetected for months. Only if you have a stand-alone system and the power switches off will you find out that you have a problem.

The faults are typically to be found in one of the following areas:

- Excessive power usage – i.e. you are using more power than you anticipated
- Insufficient power generation – i.e. you are not generating as much power as you expected
- Damaged wiring/ poor connections
- Weak batteries
- Obstructions (shading)
- Faulty earth (ground)
- Inverter faults

Obstructions are a big subject by themselves, and I cover this in much more detail in Appendix A, starting on page 153. Other faults are covered below.

Excessive power usage

This is the most common reason for solar electric systems failing: the original investigations underestimated the amount of power that was required.

Almost all solar controllers provide basic information on an LCD screen that allows you to see how much power you have generated compared to how much energy you are using, and shows the amount of charge currently stored in the battery bank. Some solar controllers include more detailed information that allows you to check on a daily basis how your power generation and power usage compares.

Using this information, you can check your power drain to see if it is higher than you originally expected.

If you have an inverter in your system, you will also need to measure this information from your inverter. Some inverters have an LCD display and can provide this information, but if your system does not provide this, you can use a plug-in watt meter to measure your power consumption over a period of time.

If your solar controller or your inverter does not provide this information, you can buy a multi-meter with data logging capabilities. These will allow you to measure the current drain from the solar controller and/or your inverter over a period of time (you would typically want to measure this over a period of a day).

Attach the multi-meter across the leads from your batteries to your solar controller and inverter. Log the information for at least 24 hours. This will allow you identify how much power is actually being used.

Some data logging multi-meters will plot a chart showing current drain at different times of the day, which can also help you identify when the drain is highest.

Solutions

If you have identified that you are using more power than you were originally anticipating, you have three choices:

- Reduce your power load
- Increase the size of your solar array
- Add another power source (such as a fuel cell, wind turbine or generator) to top up your solar electric system when necessary

Insufficient power generation

If you have done your homework correctly, you should not have a problem with insufficient power generation when the system is relatively new.

However, over a period of a few years, the solar panels and batteries will degrade in their performance (batteries more so than the solar panels), whilst new obstructions that cut out sunlight may now be causing problems.

You may also be suffering with excessive dirt on the solar panels themselves, which can significantly reduce the amount of energy the solar array can generate. Pigeons and cats are the worst culprits for this!

Your site may have a new obstruction that is blocking sunlight at a certain time of day: a tree that has grown substantially since you carried out the original site survey, for instance.

Alternatively, you may have made a mistake with the original site survey and not identified an obstruction. Unfortunately, this is the most common mistake made by inexperienced solar installers. It is also the most expensive problem to fix. This is why carrying out the site survey is so important.

To identify if your system is not generating as much power as originally expected, check the input readings on your solar controller to see how much power has been generated by your solar panels on a daily basis. If your solar controller cannot provide this information, use a multi-meter with data logger to record the amount of energy captured by the solar panels over a three-to-five day period.

Solutions

If you have identified that you are not generating as much power as you should be, start by checking your solar array. Check for damage on the solar array and then give the array a good wash with warm, soapy water and polish using a water- and dirt-repellent glass polish or wax.

Check all the wiring. Make sure that there is no unexplained high resistance in any of the solar panels or on any run of wiring. It could be a faulty connection or a damaged cable that is causing the problems.

Carry out another site survey and ensure there are no obstructions between the solar array and the sun. Double-check that the array itself is in the right position to capture the sun at solar noon. Finally, check that the array is at the optimum angle to collect sunlight.

If you are experiencing these problems only at a certain time of the year, it is worth adjusting the angle of the solar panel to provide the maximum potential power generation during this time, even if this means compromising power output at other times of year.

Check the voltage at the solar array using a multi-meter. Then check again at the solar controller. If there is a significant voltage drop between the two, the resistance in your cable is too high and you are losing significant efficiency as a result. This could be due to an inadequate cable installed in the first place, or damage in the cable. If possible, reduce the length of the cable and test again. Alternatively, replace the cable with a larger and better quality cable.

If none of that works, you have three choices:

- Reduce your power load
- Increase the size of your solar array
- Add another power source (such as a fuel cell, wind turbine or generator) to top up your solar electric system when necessary

Damaged wiring/ poor connections

If you have damaged wiring or a poor connection, this can have some very strange effects on your system. If you have odd symptoms that do not seem to add up to anything in particular, then wiring problems or poor connections are your most likely culprit.

Examples of some of the symptoms of a loose connection or damaged wiring are:

- A sudden drop in solar energy in very warm or very cold weather. This is often due to a loose connection or damaged wiring in the solar array or between the solar array and the solar controller
- Sudden or intermittent loss of power when you are running high loads. This suggests a loose connection between batteries, or between the batteries and solar controller or inverter

- Sudden or intermittent loss of power on particularly warm days after the solar array has been in the sun for a period of time. This suggests a loose connection somewhere in the array, a damaged panel or high resistance in a cable
- Significantly lower levels of power generation from the solar array suggest a loose wire connection or a short circuit between solar panels within the array
- A significant voltage drop on the cable between the solar array and the solar controller suggests either an inadequate cable or damage to the cable itself
- Likewise, a significant voltage drop on the cable between the solar controller and your low-voltage devices suggests an inadequate cable or damage to the cable itself
- If you find a cable that is very warm to the touch, it suggests the internal resistance in that cable is high. The cable should be replaced immediately

Unfortunately, diagnosing exactly where the fault is can be time-consuming. You will require a multi-meter, a test light and plenty of time.

Your first task is to identify which part of the system is failing. A solar controller that can tell you inputs and outputs is useful here. The information from this will tell you whether your solar array is underperforming or the devices are just not getting the power they need.

Once you know which part of the system to concentrate on, measure the resistance of each cable using the ohm setting on your multi-meter. If the internal resistance is higher than you would expect, replace it. If any cable is excessively hot, replace it. The problem could be caused either by having an inadequate cable in the first place (i.e. too small) or by internal damage to the cable.

Next, check all the connections in the part of the system you are looking at. Make sure the quality of the connections is good. Make sure that all cables are terminated with proper terminators or soldered. Make sure there is no water ingress.

Weak battery

The symptoms of a weak battery are that either the system does not give you as much power as you need, or you get intermittent power failures when you switch on a device.

In extreme cases, a faulty battery can actually reverse its polarity and pull down the efficiency of the entire bank.

Weak battery problems first show themselves in cold weather and when the batteries are discharged to below 50–60% capacity. In warm weather, or when the batteries are charged up, weak batteries can quite often continue to give good service for many months or years.

If your solar controller shows that you are getting enough power in from your solar array to cope with your loads, then your most likely suspect is a weak battery within your battery bank, or a bad connection between two batteries.

Start with the cheap and easy stuff. Clean all your battery terminals, check your battery interconnection cables, make sure the cable terminators are fitting tightly on the batteries and coat each terminal with a layer of petroleum jelly in order to ensure good connectivity and protection from water ingress.

Then check the water levels in your batteries (if they are 'wet' batteries). Top up as necessary.

Check to make sure that each battery in your battery bank is showing a similar voltage. If there is a disparity of more than 0.7 volts, it suggests that you may need to balance your batteries.

If, however, you are seeing a disparity on one battery of 2 volts or over, it is likely that you have a failed cell within that battery. You will probably find that this battery is also abnormally hot. Replace that battery immediately.

If your solar controller has the facility to balance batteries, then use this. If not, top up the charge on the weaker batteries, using an appropriate battery charger, until all batteries are reading a similar voltage.

If you are still experiencing problems after carrying out these tests, you will need to run a load test on all your batteries in turn. To do this, make sure all your batteries are fully charged up, disconnect the batteries from each other and use a battery load tester (you can hire these cheaply from tool hire companies). This load tester will identify any weak batteries within your bank.

Changing batteries

If all your batteries are several years old and you believe they are getting to the end of their useful lives, it is probably worth replacing the whole battery bank in one go. Badly worn batteries and new batteries do not necessarily mix well, because of the voltage difference. If you mix new and used, you can easily end up with a bank where some of the batteries never fully charge up.

If you have a bank of part-worn batteries and one battery has failed prematurely, it may be worth finding a second-hand battery of the same make and model as yours. Many battery suppliers can supply you with second-hand batteries: not only are they much cheaper than new, but because the second-hand battery will also be worn, it will have similar charging and discharging characteristics to your existing bank, which can help it bed down into your system.

If you cannot find a part-worn battery, you can use a new one, but make sure you use the same make and model as the other batteries in your bank. Never mix and match different models of batteries, as they all have slightly different characteristics.

If you add a new battery to a part-worn bank, you may find the life of the new battery is less than you would expect if you replaced all them. Over a few months of use, the performance of the new battery is likely to degrade to similar levels to the other batteries in the bank.

Before changing your battery, make sure that all of your batteries (both new and old) are fully charged.

Put a label on the new battery, noting the date it was changed. This will come in useful in future years when testing and replacing batteries.

Once you have replaced your battery, take your old one to your local scrap merchants. Lead acid batteries have a good scrap value and they can be 100% recycled to make new batteries.

Inverter issues

The symptoms of inverter issues can include:

- Buzzing or humming sounds from some electronic equipment when powered from the inverter
- Failure of some equipment to run from the inverter
- Regular tripping of circuits
- Sudden loss of power

If you are experiencing buzzing or humming sounds from electronic equipment when powered from the inverter, or if some equipment is not running at all, it suggests that the inverter is not producing a pure AC sine wave. If you have a grid-tie system, the AC pure sine waveform generated by the inverter may not be perfectly coordinated with the waveform from the grid. This would suggest poor quality power from the grid, a grounding issue or a faulty inverter.

If you have a stand-alone system and have purchased a modified sine wave inverter (or quasi-sine wave), it may be that you cannot resolve these issues without replacing the inverter. Some electronic equipment, such as laptop computers and portable televisions, may not work at all using a modified sine wave inverter, whilst other equipment will emit a buzz when run from these inverters.

If you have sudden and unexplained tripping of circuits when running from an inverter, or experience sudden loss of power, there are a number of things to check:

- Does the tripping occur when a heavy-load appliance such as a fridge switches itself on or off?
- Does the tripping occur when the inverter cuts in at the start of the day or when it cuts out at the end of the day?
- Does the tripping occur more often on very warm days or after heavy rain?

Unfortunately, circuit tripping and sudden power loss often only occurs when a combination of events occur, which can make diagnosis time-consuming and difficult to get right.

The most common reasons for circuit tripping or sudden power loss are temperature-related issues. Inverters can generate a huge amount of heat. The hotter they get, the less power they produce. If the inverter is running too hot, a sudden peak demand can be enough to shut the inverter down momentarily. If the inverter runs too hot for too long, it will shut down for a longer period of time in order to cool.

If this is the case, you are going to have to provide your inverter with more ventilation. If you cannot keep it cool, it may also mean that you require a more powerful inverter in order to cope with the load.

If the issue occurs during sudden rain or on very hot days, you may also have a grounding problem. Check the inverter with a PAT tester to ensure that you are not getting a ground leakage from the inverter itself. If you are, check all the connections from the DC input of your inverter.

Maintaining Your System

There is very little maintenance to be carried out on a solar electric system. There are some basic checks that should be carried out on a regular basis. Typically these should take no more than a few minutes to carry out.

As required

- Clean the solar array. This actually takes very little effort: unless you live in a very dry and dusty part of the world, the rain will usually do a very good job of washing your solar panels clean on a regular basis
- Telescopic window cleaner kits are available to clean solar arrays mounted on lower sections of a roof. If you can easily access the panels, a dirt- and rain-repellent glass polish can help keep your solar array cleaner for longer
- If you have a thick layer of snow on your solar array, brush it off! A thick blanket of snow very quickly stops your solar array from producing any energy at all

Every month

- If your solar controller or inverter includes a display that shows power input and power output, check the performance of your solar array. Check that it is in line with your expectations for the time of year. It is worth keeping a log of the performance so you can compare it from one year to the next
- If there is an unexplained drop in performance, clean the solar array, visually check the condition of the cables and balance the batteries. If the performance does not improve, follow the troubleshooting guide for further assistance

Every three months

- Check the ventilation in the battery box
- Check the battery area is still weatherproof and there are no leaks from the batteries
- Clean dirt and dust off the top of batteries
- Visually check all the battery connectors. Make sure they are tightly-fitting. Clean and protect them with petroleum jelly where required
- Check the electrolyte level in batteries and top up with distilled water where required

Every six months

- If you have a multi-battery system and your solar controller has the facilities to do so, balance the battery bank
- Once the batteries are balanced, use a volt meter or multi-meter to check the voltage on each individual battery. Ensure the voltages are within 0.7 volts of each other

- If one or more battery has a big difference in voltage, follow the instructions on weak batteries in the troubleshooting section of this book

Every year

- If you have a battery bank with more than two batteries, swap the order of the batteries in your bank. Place the batteries that were in the middle of your bank at either end and the batteries that were at either end of your bank in the middle. Then balance them
- This will ensure that all the batteries get even wear throughout their lifetime and thereby increase the overall lifespan of the batteries

At the start of each winter

- Check the insulation around the batteries
- Check that the area around the batteries is free of rodents. Mice and rats like to keep warm, and insulation around batteries is a tempting target. If they have found your batteries, they are likely to gnaw the cabling as well
- Clean the solar array to ensure you get the best possible performance at the worst time of the year

Internet Support

A free website supports this book. It provides up-to-the-minute information and online solar energy calculators to help simplify the cost analysis and design of your solar electric system.

To visit this site, go to the following address:

www.SolarElectricityHandbook.com

Tools available on the website

Online project analysis

The online project analysis tool on this site takes away a lot of the calculations that are involved with designing a new solar electric system, including estimating the size and type of solar panel, the size and type of battery, the thickness of low-voltage cable required and providing cost and timescale estimates.

To use the online project analysis, you will need to have completed your power analysis (see the chapter on Project Scoping), and ideally completed your whole project scope. The solar calculator will factor in the system inefficiencies and produce a thirteen-page analysis for your project.

Monthly insolation figures

Monthly solar insolation figures for every country in the world are included on the website. These can be accessed by selecting your country and the name of your nearest town or city from a list. Every country in the world is included.

The solar insolation figures used are monthly averages based on three-hourly samples taken over a 22-year period.

Solar angle calculator

The solar angle calculator shows the optimum angle for your solar panel on a month-by-month basis, and shows where the sun will rise and set at different times of the year.

Solar resources

A directory of solar suppliers is included on the website, along with links for finding out the very latest about grant schemes and selling your electricity back to the electricity companies.

Questions and answers

The site includes an extensive list of questions posted on the site by other site visitors, along with my answers. These questions and answers cover almost every conceivable area of solar design and installation and are worth a browse. If you have a question of your own, post it on the site.

Author online!

If you have noticed a mistake in the book, or feel a topic has not been covered in enough detail, I would welcome your feedback.

This handbook is updated on a yearly basis and any suggestions you have for the next edition would be gratefully received.

The website also includes an 'ask me a question' facility, so you can get in touch with any other questions you may have, or simply browse through the questions and answers in the Frequently Asked Questions section.

Solar articles

New articles about solar power are regularly added to our articles section.

A Final Word

Solar electric power is an excellent and practical resource. It can be harnessed relatively easily and effectively.

It is not without its drawbacks and it is not suitable for every application. To get the best out of a solar electric system, it is important to do your planning first and to be meticulous with detail. Only then will you have a system that will perform properly.

From an enthusiast's perspective, designing and building a solar electric system from scratch is interesting, educational and fun. If you are tempted to have a go, start with something small like a shed light, and feel free to experiment with different ideas.

If you are a professional architect or builder, you should now have a clear idea of how solar energy can be used in your projects: its benefits and its drawbacks.

It is quite amazing, the first time you connect a solar panel up to an electrical item such as a light bulb and watch it power up straight away. Even though you *know* it is going to work, there is something almost magical about watching a system that generates electricity seemingly from thin air!

Solar is making the seemingly impossible possible for the first time. In 2011 and 2012, I was involved in a project to provide solar power to remote villages in Africa, providing a reliable source of electricity for lighting, communications and the ability to store medication safely at chilled temperatures. We installed systems in Ethiopia, Kenya, Uganda and the Sudan. Total cost for all the equipment for an entire village wide system is less than $2,000. It is a tiny cost for revolutionizing an entire village.

This past year, my current solar project has been creating a solar electric delivery van for use in big cities. The vehicle is so efficient it can provide a solar-only range of around 15 miles (25km) in dull overcast winter conditions in London, or up to over 100 miles of solar range in California on a summer day. Prototypes are now running in the United Kingdom and the vehicle goes on general sale this Spring. For its chosen role, we expect our solar delivery van will be entirely solar powered and will never need to be plugged in to charge up.

Teaching children about solar electricity is also fun. There are small solar powered kits suitable for children. These can be assembled by little fingers and they teach the fundamentals of electricity and solar power in a fun and interesting way.

If this book has inspired you to install a solar electric system yourself, then it has served its purpose. I wish you the very best for your project.

All the best

Michael Boxwell
January 2015

Appendix A – Crystalline Solar Panels and Shading

As mentioned a few times in this book, shading is a big issue with crystalline photovoltaic panels. A small amount of shading has a big impact on the amount of electrical power generated by your solar system.

Unlike solar thermal (hot water) systems, the loss of power through shading is much greater than the amount of the panel that is in shade. With solar thermal systems, if only 5% of the panel is in shade, you lose less than 5% of the power production. With amorphous (thin-film) PV panels, you would usually lose 15–20% of the power production in a similar scenario.

With crystalline solar panels, the difference is significantly more and in some instances can bring power generation down to zero, even if the amount of the panel in shade is small.

The reason for this is in the construction of the solar panel itself. A crystalline solar panel is made up of a number of individual solar cells. Typically, each of these cells generates ½ volt of potential energy. These cells are connected together in series to increase the voltage to a more useful level inside the solar panel. Each individual solar panel has one or more strings of solar cells.

Because these strings of cells are connected in series, a solar panel is only as good as its weakest cell. If one cell produces a weak output, all the other cells within the string are compromised as well. This means that if you have a 'soft shade' such as a distant tree branch, creating a soft dappled shade across just one or two cells on a solar panel, the effect can be to reduce the output of the whole string to a similar level to a dull, overcast day.

Worse happens when you have a more direct shadow, creating a bigger contrast between light and shade. When even one cell is in complete shade and the remainder are in bright sunlight, the shaded cell short-circuits as the flow of electrons within the cell goes into reverse.

However, because the cell is connected in series with the other cells in the panel, current is forced through the reversed cell. In this scenario, the reversed cell *absorbs* the power produced by the other cells in the panel, generating heat and creating a hotspot within the solar panel.

The amount of power that is absorbed by a single reversed cell is disproportionate to the amount of power a single cell can generate. A single cell may only produce ½ volt of potential energy, but can absorb 6–8 volts when it has gone into reversal.

If unchecked, a solar cell left in this state can easily be destroyed. The hot spot generated by the blocked cell can very quickly reach dangerous temperature levels too, and amateur home-built solar panels have been known to burst into flames precisely for this reason.

Thankfully, solar panel manufacturers have a solution to this problem in order to avoid the panel itself becoming damaged. All professionally manufactured crystalline solar panels have in-built protection to route power around a string of cells where one or more of the cells are in reversal.

If your solar panel has only a single string of solar cells within it, this effectively means the entire solar panel is bypassed and produces no power at all. If your solar panel has two strings of solar cells, it means the power output of the panel is halved.

Because a solar array is typically put together by installing multiple solar panels in series, the effects of shading on just part of one panel impacts the performance of the entire solar array.

Types of obstruction

Shade can be broken down into two categories: soft shade and hard shade.

A soft shade is a distant obstruction where the shadow is dispersed or diffused, which significantly *reduces* the amount of light reaching the solar cells. A shadow from a tree would be classed as a soft shade.

Hard shade is an obstruction that blocks out light from reaching the solar cell completely. Bird droppings, fallen leaves or a tree branch sitting on top of the glass would be classed as a hard shade.

Where cells are soft shaded, you will see a significant drop off in energy production. Effectively, the production of the entire array will drop down to a similar level to a dull day.

Where cells are hard shaded, the production of the entire array will drop down to the same level as the affected cells: if the cells are covered completely, you may see a complete power shutdown (depending on how shade-tolerant your solar panels are). If the cells are only partially covered, you will see a significant drop in energy production.

Designing shade-tolerant solar systems

If you cannot avoid shade in your solar energy system, the solution is to design a *shade-tolerant* system. In other words, you have to design your system in such a way that the effect of shade on any one part of your system has as small an effect on the overall array as possible.

Designing shade-tolerant solar arrays is a complex subject and a specialist area even amongst solar design experts. There are entire books written on this subject alone. Consequently, it is not possible to cover the whole subject here. However, it is possible to design a basic solar energy system with a reasonable level of shade tolerance without a huge amount of specialist knowledge.

If you need to design a system that will continue to perform well in partially shaded conditions, there are options available to you:

Track the shade

First, you should never design a system that will have to cope with hard shading. Solar panels should be mounted so that they are not covered in leaves and so they can be inspected and cleaned of any hard obstructions if necessary. If there is a permanent obstruction that will be creating a hard shade, you should not be installing solar panels in that particular location.

If you have a soft shade, for how much of the day does this shade your panels? Remember, even a soft shade over a small area can have a big impact. Typical core power generation times for solar energy during the summer are three hours either side of solar noon (i.e. between 9am

154

and 3pm, if your time zone equates to solar time). If you have shading either before this period or after it, you will lose around 20% of your capability in the summer, or 40% of your capability if you have shading in both the early morning and late afternoon.

During the winter, the difference is not so great. Because the sun is lower in the sky and the intensity of the sunlight is significantly lower, almost all of your power is generated during the core power generation times. If you have shading before 9am or after 3pm during the winter, you will probably be losing only around 5–10% of your generating capabilities.

However, if you are suffering from shade within the core power generation times, you are going to severely compromise the performance of your system. The exact impact on the performance of your system will vary according to your location, the severity of shading and the type of solar panels you are using, but in general terms, this is how much of your solar energy production you can expect to lose from one hour of shading during core power generation times:

Period of Shading	Winter loss (%)	Summer loss (%)
9am–10am *(2–3 hours before solar noon)*	3%	6%
10am–11am *(1–2 hours before solar noon)*	15%	10%
11am–solar noon	27%	14%
Solar noon–1pm	27%	14%
1pm–2pm *(1–2 hours after solar noon)*	15%	10%
2pm–3pm *(2–3 hours after solar noon)*	3%	6%

This table shows the approximate performance loss from your system if it is shaded for one hour during core production times

As you see, you can lose a significant amount of energy production due to shading during the middle of the day. This is why shading during the middle of the day is the number one reason for solar arrays failing to live up to expectations.

However, because you can now quantify the impact of the shading, it is possible to do something to counter the effects.

Increasing the number of solar panels

The first option is the most obvious: if you have the space, you can increase the number of solar panels you install in order to counter the periods of shade.

This is not always possible, either because of space or cost restrictions. Also, it is not always the most efficient way of getting around the problem.

Panel orientation

If shade affects you at a particular time of the day, consider angling your solar panels away from the obstruction. This will increase their effectiveness during the unobstructed parts of the day and reduce or remove the impact of the obstruction in the first place. The reduction in power generation from angling the panels away from due south is often less than the impact on shading if you can eliminate the shade problem altogether.

Choice of solar panel

Another option is to choose amorphous (thin-film) solar panels. Amorphous solar panels do not suffer from cell reversal in the same way that crystalline solar panels do, and consequently provide far better shade tolerance.

Because of their lower efficiency levels, you will need to take into account that you will require around twice as much physical space to install amorphous solar panels. If space is not an issue, using amorphous solar panels is likely to be the simplest and most cost-effective solution for *stand-alone* solar installations.

Sharp, Mitsubishi, Uni-Solar, Solar Frontier and Sanyo now manufacture high-quality amorphous solar panels that offer excellent performance and reliability.

Use micro-inverters

If you are building a grid-tie system, possibly the best solution is to use micro-inverters, where each panel has its own power inverter and is a full solar energy system in its own right.

With a micro-inverter system, each solar panel runs entirely independently of every other solar panel. If shade affects one panel, none of the other panels is affected in any way.

Using micro-inverters also means that you can have solar panels facing in different directions from each other. For instance, if you are installing solar panels on a roof and your roof has multiple pitches and angles, you can choose to install your solar array on more than one part of the roof using a micro-inverter system.

This approach gives you greater flexibility as to where you mount solar panels: it may even be possible to avoid your shading issues altogether.

Design a parallel solar array

By connecting solar panels in parallel rather than in series, you can reduce the overall effect of shading on just one or two parts of the array. In a parallel system, when one solar panel is in shade, the power outputs of other panels in the array are not affected.

If you are designing a grid-tie solar energy system and want a high-voltage system, it is possible to achieve this with a parallel array by using high-voltage grid-tie specific solar panels. A number of manufacturers now offer these panels, often producing over 100 volts from a single panel. Other panels are available that provide either 24-volt or 48-volt output, and these can be suitable for larger stand-alone solar energy systems that require more than 12 volts.

Design a multi-string solar array

Instead of designing your system to have just one set of solar panels connected in series, you can design your system to have multiple series of solar panels. Controllers and inverters are available that allow you to have multiple strings of solar panels, or you can have a controller or inverter for each individual string.

In effect, this means you end up with two smaller solar power systems, rather than one. This may mean you can face your two different sets of solar arrays at different angles, giving you the opportunity to mount them in entirely different locations if you so wish.

In this scenario, you can design your system so that partial shading will only affect one of your strings rather than your entire array. As with micro-inverters, such an approach may even allow you to avoid the shading issue altogether.

Other options

There are other options, but the designs can become extremely complicated and are really in the domain of highly specialist solar designers. If you are designing a shade-tolerant solar energy system and the above options will not work for you, it is time to call in a solar shading specialist.

Unless you have significant shading issues, you can usually design around the problems using one of the options I have suggested here, or by using a combination of options to come up with a workable solution. Over the past few years, I have designed a number of shade-tolerant systems, including systems that are designed to work in an entirely shaded environment all year round. With careful planning, it is often possible to overcome the issues.

If all else fails...

Sometimes it is not possible to design around shading problems. In this scenario, it requires a rethink of what you can achieve using solar.

This is an issue that I have with my own home. My house is on the edge of ancient woodland. For most of the year, my home is almost entirely shaded by the tall trees that surround it. Even in the height of summer, the sun does not reach my garden until mid-afternoon.

This shading means I could never run my own home completely from solar power. However, I have designed a smaller solar energy system that provides me with enough energy to run my lighting and provide backup power in case of a power cut.

I have achieved this by installing amorphous solar panels onto the back of my home, facing south-west to catch the afternoon and evening sun. Despite receiving very little sunlight, this system provides enough electricity to provide me with my lighting and backup power requirements throughout the year, even in the depths of winter.

In conclusion

- Shade is a big problem for solar energy – especially if using crystalline solar panels
- Even a very small amount of shade can have a big impact on your system
- Amorphous solar panels also suffer in shade, but not to the same extent as crystalline panels
- If you have shade during core power generating periods of the day – typically the three hours either side of solar noon – you will lose a very significant amount of your potential power generation
- There are things you can do to reduce the impact of shading, either through your choice of materials or your system design

- In some cases where you have very significant shading issues, you may need to reconsider what you can realistically achieve with solar

Appendix B – Solar Insolation

Solar insolation shows the daily amount of energy from the sun you can expect per square metre at your location, averaged out over the period of a month. The figures are presented as an average *irradiance*, measured in kilowatt-hours per square metre spread over the period of a day (kWh/m²/day).

Averages have been collated over a 22-year period between 1983 and 2005, based on a three-hour sample rate.

The book only shows solar insolation figures for the United States, Canada, Australia, New Zealand, UK and Ireland. *www.SolarElectricityHandbook.com* has figures for every major town and city in every country in the world, allowing you to access information that is specific to your area

Understanding this information

As described on page 37, the amount of energy you capture from the sun differs, depending on the tilt of the solar panels. If you mount your solar panels horizontally or vertically, you will capture less energy than if you face them due south (due north in the Southern Hemisphere) and tilt them towards the sun.

The tables on the following pages show the irradiance figures based on mounting your solar panels at the following angles:

- Flat (horizontal)
- Upright
- Tilted towards the equator for best year-round performance
- Tilted for best performance during the winter months
- Tilted for best performance during the summer months
- Tilted with the angle adjusted each month throughout the year

Where the figures show the panels tilted at a fixed angle, I show this angle in the left hand column as the angle adjustment from an upright (vertical) position. On the bottom row, where the optimum tilt changes each month, I show this angle underneath each month's irradiance figures. *Please note: all angles are in degrees from vertical.*

For a more detailed explanation of these figures, refer to the chapter on Calculating Solar Energy starting on page 37.

Solar insolation values – Australia

New South Wales

	Jan	Feb	Mar	Apr	May	Jun	Jul	Aug	Sep	Oct	Nov	Dec
Flat - 90°	5.91	5.25	4.48	3.56	2.73	2.49	2.68	3.49	4.6	5.41	5.88	6.24
Upright - 0°	2.34	2.56	2.93	3.36	3.43	3.77	3.78	3.83	3.54	2.83	2.40	2.31
56° angle Year-round tilt	5.38	5.11	4.84	4.42	3.87	3.90	4.04	4.68	5.33	5.50	5.43	5.57
40° angle Best winter tilt	4.74	4.66	4.62	4.45	4.07	4.20	4.32	4.82	5.22	5.11	4.82	4.85
72° angle Best summer tilt	5.78	5.32	4.82	4.16	3.46	3.37	3.54	4.28	5.16	5.63	5.80	6.04
Tilt adjusted each month	5.78 72°	5.34 64°	4.84 56°	4.47 48°	4.07 40°	4.27 32°	4.32 40°	4.82 48°	5.33 56°	5.63 64°	5.80 72°	6.24 80°

Northern Territory

	Jan	Feb	Mar	Apr	May	Jun	Jul	Aug	Sep	Oct	Nov	Dec
Flat - 90°	5.83	5.18	5.07	4.81	4.35	4.24	4.47	5.07	5.98	6.59	6.5	6.27
Upright - 0°	2.43	1.81	2.28	3.09	3.73	4.24	4.23	3.75	2.99	2.02	2.32	2.79
73° angle Year-round tilt	5.85	5.04	5.13	5.15	4.94	4.99	5.20	5.62	6.23	6.48	6.48	6.36
57° angle Best winter tilt	5.58	4.68	4.94	5.21	5.23	5.43	5.61	5.83	6.13	6.04	6.12	6.13
89° angle Best summer tilt	5.84	5.18	5.08	4.84	4.40	4.29	4.52	5.12	6.00	6.59	6.51	6.28
Tilt adjusted each month	5.84 89°	5.18 81°	5.13 73°	5.22 65°	5.23 57°	5.56 50°	5.61 57°	5.83 65°	6.23 73°	6.60 81°	6.51 89°	6.37 96°

Queensland

	Jan	Feb	Mar	Apr	May	Jun	Jul	Aug	Sep	Oct	Nov	Dec
Flat - 90°	6.19	5.39	4.95	3.98	3.23	3.02	3.22	4.04	5.12	5.52	6.07	6.35
Upright - 0°	2.09	2.28	2.81	3.22	3.41	3.75	3.76	3.85	3.39	2.49	2.13	2.04
62° angle Year-round tilt	5.65	5.22	5.19	4.61	4.12	4.13	4.29	4.99	5.65	5.50	5.61	5.71
46° angle Best winter tilt	4.98	4.78	4.96	4.63	4.32	4.44	4.57	5.15	5.54	5.11	4.99	4.98
78° angle Best summer tilt	6.06	5.41	5.15	4.34	3.71	3.59	3.78	4.56	5.47	5.62	5.98	6.18
Tilt adjusted each month	6.06 78°	5.42 70°	5.19 62°	4.65 54°	4.32 46°	4.50 38°	4.57 46°	5.15 54°	5.65 62°	5.62 70°	5.98 78°	6.35 86°

South Australia

	Jan	Feb	Mar	Apr	May	Jun	Jul	Aug	Sep	Oct	Nov	Dec
Flat - 90°	6.81	6.18	4.91	3.75	2.71	2.29	2.53	3.21	4.32	5.34	6.29	6.67
Upright - 0°	2.54	2.94	3.32	3.72	3.54	3.48	3.63	3.52	3.37	2.87	2.54	2.42
55° angle Year-round tilt	6.15	6.05	5.42	4.80	3.94	3.60	3.86	4.29	5.01	5.45	5.80	5.93
39° angle Best winter tilt	5.36	5.49	5.19	4.86	4.15	3.87	4.12	4.41	4.89	5.05	5.12	5.13
71° angle Best summer tilt	6.65	6.31	5.37	4.49	3.51	3.13	3.39	3.95	4.86	5.58	6.21	6.45
Tilt adjusted each month	6.65 71°	6.32 63°	5.42 55°	4.87 47°	4.15 39°	3.93 32°	4.12 39°	4.41 47°	5.01 55°	5.58 63°	6.21 71°	6.68 78°

Victoria

	Jan	Feb	Mar	Apr	May	Jun	Jul	Aug	Sep	Oct	Nov	Dec
Flat - 90°	6.36	5.83	4.51	3.23	2.23	1.78	1.94	2.59	3.55	4.72	5.74	6.22
Upright - 0°	2.62	3.00	3.24	3.35	3.02	2.77	2.81	2.89	2.85	2.72	2.56	2.49
52° angle Year-round tilt	5.75	5.74	5.04	4.22	3.33	2.87	3.00	3.48	4.10	4.80	5.30	5.53
36° angle Best winter tilt	5.02	5.20	4.81	4.25	3.48	3.06	3.17	3.55	3.98	4.43	4.69	4.79
68° angle Best summer tilt	6.22	6.00	5.01	3.97	3.00	2.52	2.67	3.24	4.02	4.94	5.67	6.01
Tilt adjusted each month	6.22 68°	6.02 60°	5.04 52°	4.27 44°	3.48 36°	3.09 28°	3.17 36°	3.55 44°	4.10 52°	4.94 60°	5.67 68°	6.23 76°

Western Australia

	Jan	Feb	Mar	Apr	May	Jun	Jul	Aug	Sep	Oct	Nov	Dec
Flat - 90°	8.41	7.49	5.93	4.34	3.09	2.62	2.82	3.62	5.04	6.41	7.71	8.45
Upright - 0°	2.46	3.08	3.76	4.12	3.77	3.68	3.71	3.74	3.74	3.09	2.55	2.3
58° angle Year-round tilt	7.51	7.31	6.53	5.46	4.29	3.89	4.05	4.69	5.80	6.54	7.06	7.39
42° angle Best winter tilt	6.47	6.61	6.28	5.55	4.53	4.18	4.32	4.83	5.69	6.07	6.16	6.28
74° angle Best summer tilt	8.18	7.62	6.43	5.07	3.82	3.37	3.56	4.3	5.59	6.67	7.59	8.14
Tilt adjusted each month	8.18 74°	7.63 66°	6.53 58°	5.56 50°	4.53 42°	4.25 34°	4.32 42°	4.83 50°	5.80 58°	6.67 66°	7.59 74°	8.45 82°

Solar insolation values – Canada

Alberta

	Jan	Feb	Mar	Apr	May	Jun	Jul	Aug	Sep	Oct	Nov	Dec
Flat - 90°	0.79	1.70	3.15	4.56	5.32	5.66	5.56	4.76	3.14	1.85	0.99	0.48
Upright - 0°	1.58	2.60	3.52	3.63	3.24	3.11	3.19	3.36	3.03	2.51	1.81	0.95
36° angle Year-round tilt	1.61	2.84	4.26	5.00	4.87	4.82	4.89	4.84	3.86	2.84	1.89	0.97
20° angle Best winter tilt	1.66	2.85	4.09	4.55	4.30	4.21	4.30	4.33	3.63	2.80	1.93	1.00
52° angle Best summer tilt	1.47	2.69	4.21	5.20	5.29	5.33	5.39	5.14	3.90	2.74	1.75	0.89
Tilt adjusted each month	1.66 20°	2.86 28°	4.26 36°	5.21 44°	5.29 52°	5.72 60°	5.39 52°	5.20 44°	3.86 36°	2.85 28°	1.93 20°	1.00 12°

British Columbia

	Jan	Feb	Mar	Apr	May	Jun	Jul	Aug	Sep	Oct	Nov	Dec
Flat - 90°	1.00	1.88	2.9	4.18	5.18	5.7	6.11	5.37	4.00	2.19	1.18	0.84
Upright - 0°	1.52	2.44	2.73	2.93	2.89	2.86	3.15	3.43	3.54	2.62	1.73	1.37
42° angle Year-round tilt	1.63	2.78	3.52	4.34	4.75	4.92	5.42	5.40	4.78	3.09	1.87	1.43
26° angle Best winter tilt	1.67	2.78	3.37	3.97	4.17	4.24	4.69	4.85	4.53	3.06	1.91	1.48
58° angle Best summer tilt	1.51	2.63	3.50	4.52	5.12	5.42	5.91	5.70	4.79	2.96	1.74	1.31
Tilt adjusted each month	1.67 26°	2.80 34°	3.52 42°	4.53 50°	5.12 58°	5.75 66°	5.91 58°	5.74 50°	4.78 42°	3.10 34°	1.91 26°	1.48 18°

Manitoba

	Jan	Feb	Mar	Apr	May	Jun	Jul	Aug	Sep	Oct	Nov	Dec
Flat - 90°	1.24	2.17	3.43	4.74	5.57	5.84	5.92	5.00	3.45	2.25	1.42	0.99
Upright - 0°	2.15	3.01	3.50	3.48	3.16	3.00	3.15	3.28	3.07	2.81	2.34	1.84
40° angle Year-round tilt	2.24	3.37	4.41	5.08	5.13	5.05	5.27	5.04	4.09	3.26	2.47	1.87
24° angle Best winter tilt	2.31	3.39	4.24	4.65	4.51	4.42	4.57	4.53	3.87	3.24	2.54	1.95
56° angle Best summer tilt	2.04	3.18	4.34	5.26	5.53	5.57	5.74	5.31	4.12	3.12	2.27	1.69
Tilt adjusted each month	2.31 24°	3.41 32°	4.41 40°	5.26 48°	5.53 56°	5.90 64°	5.74 56°	5.36 48°	4.09 40°	3.27 32°	2.54 24°	1.95 16°

New Brunswick

	Jan	Feb	Mar	Apr	May	Jun	Jul	Aug	Sep	Oct	Nov	Dec
Flat - 90°	1.61	2.46	3.68	4.46	4.91	5.45	5.25	4.78	3.75	2.45	1.53	1.28
Upright - 0°	2.53	3.11	3.45	3.03	2.64	2.64	2.66	2.90	3.10	2.81	2.19	2.06
45° angle Year-round tilt	2.67	3.59	4.52	4.69	4.56	4.75	4.68	4.73	4.32	3.37	2.39	2.15
29° angle Best winter tilt	2.77	3.61	4.37	4.30	4.03	4.11	4.09	4.27	4.1	3.35	2.45	2.24
61° angle Best summer tilt	2.43	3.37	4.45	4.84	4.90	5.21	5.08	4.98	4.34	3.22	2.20	1.95
Tilt adjusted each month	2.77 29°	3.62 37°	4.52 45°	4.85 53°	4.90 61°	5.48 68°	5.08 61°	5.02 53°	4.32 45°	3.38 37°	2.45 29°	2.24 22°

Newfoundland and Labrador

	Jan	Feb	Mar	Apr	May	Jun	Jul	Aug	Sep	Oct	Nov	Dec
Flat - 90°	1.23	2.07	3.27	4.16	4.75	5.16	4.90	4.36	3.25	2.09	1.26	0.99
Upright - 0°	1.91	2.68	3.12	2.87	2.64	2.62	2.59	2.73	2.72	2.43	1.83	1.63
43° angle Year-round tilt	2.03	3.06	4.03	4.30	4.34	4.48	4.35	4.29	3.73	2.89	1.99	1.7
27° angle Best winter tilt	2.09	3.07	3.88	3.94	3.83	3.88	3.80	3.87	3.53	2.86	2.03	1.77
59° angle Best summer tilt	1.87	2.89	3.99	4.47	4.68	4.93	4.74	4.53	3.77	2.77	1.85	1.55
Tilt adjusted each month	2.09 27°	3.08 35°	4.03 43°	4.48 51°	4.68 59°	5.20 66°	4.74 59°	4.58 51°	3.73 43°	2.90 35°	2.03 27°	1.77 20°

Nova Scotia

	Jan	Feb	Mar	Apr	May	Jun	Jul	Aug	Sep	Oct	Nov	Dec
Flat - 90°	1.54	2.39	3.47	4.29	4.98	5.57	5.49	4.85	3.91	2.66	1.57	1.21
Upright - 0°	2.85	3.27	3.34	2.84	2.59	2.59	2.66	2.87	3.28	3.23	2.49	2.25
45° angle Year-round tilt	2.89	3.70	4.39	4.45	4.62	4.85	4.90	4.82	4.62	3.82	2.64	2.27
29° angle Best winter tilt	3.06	3.77	4.26	4.09	4.07	4.18	4.26	4.34	4.40	3.84	2.74	2.4
61° angle Best summer tilt	2.56	3.42	4.28	4.61	4.96	5.31	5.32	5.06	4.60	3.60	2.39	2.01
Tilt adjusted each month	3.06 29°	3.78 37°	4.39 45°	4.61 53°	4.96 61°	5.60 68°	5.32 61°	5.10 53°	4.62 45°	3.86 37°	2.74 29°	2.42 22°

162

Ontario

	Jan	Feb	Mar	Apr	May	Jun	Jul	Aug	Sep	Oct	Nov	Dec
Flat - 90°	1.53	2.35	3.29	4.35	5.12	5.88	5.87	5.02	3.92	2.64	1.55	1.24
Upright - 0°	2.63	3.04	3.02	2.83	2.61	2.64	2.74	2.91	3.21	3.07	2.29	2.21
46° angle Year-round tilt	2.71	3.50	4.04	4.50	4.75	5.12	5.24	4.98	4.58	3.68	2.47	2.24
30° angle Best winter tilt	2.85	3.55	3.91	4.13	4.19	4.41	4.55	4.49	4.36	3.69	2.56	2.37
62° angle Best summer tilt	2.41	3.25	3.96	4.65	5.10	5.6	5.69	5.23	4.56	3.48	2.25	1.99
Tilt adjusted each month	2.85 30°	3.56 38°	4.04 46°	4.66 54°	5.10 62°	5.90 70°	5.69 62°	5.27 54°	4.58 46°	3.71 38°	2.56 30°	2.39 22°

Prince Edward Island

	Jan	Feb	Mar	Apr	May	Jun	Jul	Aug	Sep	Oct	Nov	Dec
Flat - 90°	1.35	2.10	3.22	4.28	5.40	6.05	6.04	5.22	3.91	2.34	1.29	0.98
Upright - 0°	2.09	2.63	2.98	2.9	2.89	2.88	3.00	3.20	3.30	2.70	1.81	1.51
44° angle Year-round tilt	2.21	3.04	3.90	4.42	5.01	5.23	5.37	5.20	4.57	3.23	1.98	1.6
28° angle Best winter tilt	2.29	3.05	3.75	4.05	4.40	4.50	4.65	4.68	4.33	3.20	2.02	1.65
60° angle Best summer tilt	2.03	2.87	3.87	4.59	5.40	5.75	5.85	5.48	4.58	3.09	1.84	1.47
Tilt adjusted each month	2.29 28°	3.06 36°	3.90 44°	4.60 52°	5.40 60°	6.09 68°	5.85 60°	5.52 52°	4.57 44°	3.24 36°	2.02 28°	1.65 20°

Quebec

	Jan	Feb	Mar	Apr	May	Jun	Jul	Aug	Sep	Oct	Nov	Dec
Flat - 90°	1.51	2.44	3.60	4.51	4.97	5.48	5.21	4.70	3.49	2.15	1.40	1.18
Upright - 0°	2.42	3.16	3.44	3.10	2.71	2.7	2.69	2.91	2.91	2.47	2.03	1.97
43° angle Year-round tilt	2.55	3.61	4.46	4.7	4.59	4.76	4.64	4.66	4.02	2.95	2.21	2.04
27° angle Best winter tilt	2.64	3.64	4.31	4.31	4.05	4.12	4.05	4.20	3.81	2.93	2.26	2.12
59° angle Best summer tilt	2.33	3.40	4.40	4.87	4.93	5.23	5.04	4.91	4.04	2.83	2.04	1.85
Tilt adjusted each month	2.64 27°	3.65 35°	4.46 43°	4.88 51°	4.93 59°	5.52 66°	5.04 59°	4.95 51°	4.02 43°	2.96 35°	2.26 27°	2.13 20°

Saskatchewan

	Jan	Feb	Mar	Apr	May	Jun	Jul	Aug	Sep	Oct	Nov	Dec
Flat - 90°	1.11	1.98	3.25	4.79	5.60	5.78	6.23	5.06	3.65	2.28	1.36	0.91
Upright - 0°	1.94	2.77	3.32	3.57	3.21	3.00	3.33	3.36	3.33	2.90	2.30	1.74
40° angle Year-round tilt	2.02	3.10	4.16	5.15	5.15	4.99	5.54	5.11	4.40	3.36	2.41	1.76
24° angle Best winter tilt	2.08	3.11	4.00	4.71	4.51	4.36	4.78	4.58	4.16	3.33	2.48	1.83
56° angle Best summer tilt	1.85	2.92	4.11	5.34	5.56	5.51	6.04	5.39	4.42	3.21	2.22	1.59
Tilt adjusted each month	2.08 24°	3.12 32°	4.16 40°	5.35 48°	5.56 56°	5.85 64°	6.04 56°	5.44 48°	4.44 40°	3.37 32°	2.48 24°	1.84 16°

Solar insolation values – Ireland

Borders Region

	Jan	Feb	Mar	Apr	May	Jun	Jul	Aug	Sep	Oct	Nov	Dec
Flat - 90°	0.57	1.25	2.34	3.96	5.34	5.41	4.92	4.10	2.87	1.53	0.71	0.4
Upright - 0°	1.06	1.85	2.44	3.13	3.31	3.04	2.89	2.90	2.77	2.05	1.23	0.81
35° angle Year-round tilt	1.10	2.03	2.98	4.25	4.89	4.61	4.31	4.10	3.50	2.32	1.29	0.82
19° angle Best winter tilt	1.12	2.02	2.83	3.87	4.33	4.03	3.79	3.66	3.29	2.28	1.31	0.84
51° angle Best summer tilt	1.01	1.94	2.98	4.44	5.32	5.09	4.75	4.36	3.55	2.25	1.21	0.75
Tilt adjusted each month	1.12 19°	2.04 27°	2.98 35°	4.46 43°	5.32 51°	5.47 58°	4.75 51°	4.43 43°	3.50 35°	2.32 27°	1.31 19°	0.84 12°

Dublin, The Midlands and Midlands-East Regions

	Jan	Feb	Mar	Apr	May	Jun	Jul	Aug	Sep	Oct	Nov	Dec
Flat - 90°	0.62	1.19	2.09	3.34	4.32	4.36	4.21	3.52	2.53	1.44	0.82	0.5
Upright - 0°	1.03	1.59	1.97	2.45	2.61	2.46	2.44	2.37	2.25	1.78	1.33	0.92
37° angle Year-round tilt	1.08	1.78	2.48	3.42	3.90	3.76	3.70	3.42	2.93	2.05	1.41	0.94
21° angle Best winter tilt	1.10	1.77	2.34	3.10	3.47	3.3	3.26	3.06	2.74	2.01	1.43	0.97
53° angle Best summer tilt	1.01	1.71	2.50	3.59	4.25	4.14	4.05	3.64	2.98	1.99	1.32	0.87
Tilt adjusted each month	1.10 21°	1.79 29°	2.48 37°	3.62 45°	4.25 53°	4.42 60°	4.05 53°	3.71 45°	2.93 37°	2.05 29°	1.43 21°	0.97 14°

South-East, West and South West Regions

	Jan	Feb	Mar	Apr	May	Jun	Jul	Aug	Sep	Oct	Nov	Dec
Flat - 90°	0.67	1.26	2.15	3.46	4.46	4.53	4.41	3.72	2.66	1.52	0.87	0.56
Upright - 0°	1.06	1.65	2.00	2.52	2.66	2.51	2.52	2.48	2.35	1.86	1.37	0.99
38° angle Year-round tilt	1.13	1.86	2.53	3.54	4.04	3.90	3.88	3.63	3.08	2.16	1.47	1.02
22° angle Best winter tilt	1.15	1.84	2.40	3.22	3.59	3.43	3.43	3.25	2.88	2.12	1.49	1.05
54° angle Best summer tilt	1.05	1.79	2.55	3.72	4.39	4.30	4.25	3.86	3.13	2.10	1.37	0.94
Tilt adjusted each month	1.15 22°	1.87 30°	2.53 38°	3.74 46°	4.39 54°	4.58 62°	4.25 54°	3.92 46°	3.08 38°	2.16 30°	1.49 22°	1.05 14°

Solar insolation values – New Zealand

North Island

	Jan	Feb	Mar	Apr	May	Jun	Jul	Aug	Sep	Oct	Nov	Dec
Flat - 90°	6.41	5.65	4.59	3.31	2.35	1.96	2.18	2.83	3.96	4.82	5.86	6.26
Upright - 0°	2.58	2.87	3.23	3.36	3.14	3.05	3.20	3.17	3.20	2.73	2.55	2.46
53° angle Year-round tilt	5.80	5.54	5.10	4.27	3.47	3.14	3.39	3.82	4.63	4.92	5.41	5.57
37° angle Best winter tilt	5.07	5.03	4.87	4.3	3.64	3.36	3.60	3.91	4.52	4.56	4.79	4.83
69° angle Best summer tilt	6.27	5.79	5.06	4.02	3.12	2.75	2.99	3.53	4.51	5.05	5.79	6.06
Tilt adjusted each month	6.27 69°	5.81 61°	5.10 53°	4.32 45°	3.64 37°	3.41 30°	3.60 37°	3.91 45°	4.63 53°	5.05 61°	5.79 69°	6.27 76°

South Island

	Jan	Feb	Mar	Apr	May	Jun	Jul	Aug	Sep	Oct	Nov	Dec
Flat - 90°	5.75	5.01	3.92	2.67	1.82	1.36	1.60	2.28	3.39	4.50	5.49	5.82
Upright - 0°	2.75	2.94	3.18	3.14	3.05	2.64	3.01	3.08	3.17	2.90	2.76	2.67
46° angle Year-round tilt	5.21	4.98	4.56	3.77	3.19	2.63	3.03	3.49	4.23	4.66	5.09	5.07
30° angle Best winter tilt	4.54	4.50	4.34	3.78	3.35	2.81	3.23	3.57	4.10	4.28	4.47	4.39
62° angle Best summer tilt	5.64	5.23	4.55	3.56	2.85	2.30	2.67	3.23	4.13	4.82	5.47	5.55
Tilt adjusted each month	5.64 62°	5.26 54°	4.56 46°	3.80 38°	3.35 30°	2.84 22°	3.23 30°	3.57 38°	4.23 46°	4.82 54°	5.47 62°	5.84 70°

Solar insolation values – United Kingdom

London

	Jan	Feb	Mar	Apr	May	Jun	Jul	Aug	Sep	Oct	Nov	Dec
Flat - 90°	0.75	1.37	2.31	3.57	4.59	4.86	4.82	4.20	2.81	1.69	0.92	0.60
Upright - 0°	1.20	1.80	2.18	2.58	2.7	2.64	2.71	2.80	2.47	2.07	1.43	1.01
38° angle Year-round tilt	1.27	2.04	2.76	3.67	4.17	4.2	4.25	4.16	3.26	2.41	1.53	1.05
22° angle Best winter tilt	1.30	2.03	2.62	3.34	3.66	3.69	3.76	3.73	3.06	2.37	1.56	1.08
54° angle Best summer tilt	1.19	1.95	2.77	3.84	4.52	4.63	4.66	4.41	3.31	2.33	1.43	0.97
Tilt adjusted each month	1.30 22°	2.05 30°	2.76 38°	3.86 46°	4.52 54°	4.91 62°	4.66 54°	4.46 46°	3.26 38°	2.41 30°	1.56 22°	1.08 14°

South East

	Jan	Feb	Mar	Apr	May	Jun	Jul	Aug	Sep	Oct	Nov	Dec
Flat - 90°	0.80	1.44	2.42	3.70	4.73	4.99	5.00	4.31	2.88	1.75	0.95	0.62
Upright - 0°	1.34	1.94	2.33	2.71	2.8	2.72	2.81	2.9	2.57	2.18	1.51	1.09
38° angle Year-round tilt	1.41	2.18	2.94	3.84	4.31	4.32	4.42	4.3	3.37	2.53	1.61	1.12
22° angle Best winter tilt	1.44	2.18	2.81	3.5	3.79	3.80	3.92	3.86	3.17	2.49	1.64	1.16
54° angle Best summer tilt	1.30	2.08	2.94	4.01	4.68	4.76	4.85	4.55	3.41	2.43	1.50	1.03
Tilt adjusted each month	1.44 22°	2.20 30°	2.94 38°	4.03 46°	4.68 54°	5.05 62°	4.85 54°	4.60 46°	3.37 38°	2.53 30°	1.64 22°	1.16 14°

South West

	Jan	Feb	Mar	Apr	May	Jun	Jul	Aug	Sep	Oct	Nov	Dec
Flat - 90°	0.81	1.51	2.49	3.91	5.13	5.37	5.28	4.37	3.07	1.74	1.01	0.65
Upright - 0°	1.26	1.98	2.36	2.83	2.97	2.84	2.9	2.88	2.72	2.09	1.54	1.08
39° angle Year-round tilt	1.34	2.25	3.00	4.07	4.7	4.64	4.68	4.34	3.60	2.45	1.65	1.13
23° angle Best winter tilt	1.37	2.24	2.86	3.72	4.13	4.08	4.07	3.90	3.40	2.41	1.68	1.16
55° angle Best summer tilt	1.25	2.14	3.00	4.24	5.08	5.12	5.12	4.59	3.64	2.36	1.54	1.04
Tilt adjusted each month	1.37 23°	2.26 31°	3.00 39°	4.26 47°	5.08 55°	5.43 62°	5.12 55°	4.65 47°	3.60 39°	2.45 31°	1.68 23°	1.16 16°

East of England

	Jan	Feb	Mar	Apr	May	Jun	Jul	Aug	Sep	Oct	Nov	Dec
Flat - 90°	0.72	1.34	2.37	3.60	4.68	4.90	4.86	4.20	2.82	1.64	0.91	0.57
Upright - 0°	1.18	1.80	2.30	2.63	2.78	2.68	2.75	2.83	2.52	2.04	1.45	1.01
38° angle Year-round tilt	1.25	2.03	2.89	3.71	4.24	4.21	4.27	4.16	3.30	2.36	1.55	1.04
22° angle Best winter tilt	1.27	2.02	2.74	3.38	3.77	3.69	3.77	3.73	3.10	2.32	1.57	1.07
54° angle Best summer tilt	1.16	1.94	2.90	3.89	4.61	4.65	4.69	4.41	3.35	2.29	1.45	0.96
Tilt adjusted each month	1.27 22°	2.04 30°	2.89 38°	3.91 46°	4.61 54°	4.96 62°	4.69 54°	4.48 46°	3.30 38°	2.37 30°	1.57 22°	1.07 14°

East Midlands

	Jan	Feb	Mar	Apr	May	Jun	Jul	Aug	Sep	Oct	Nov	Dec
Flat - 90°	0.64	1.31	2.20	3.37	4.44	4.51	4.50	3.82	2.59	1.54	0.8	0.51
Upright - 0°	1.07	1.80	2.11	2.47	2.67	2.53	2.59	2.59	2.31	1.93	1.27	0.94
37° angle Year-round tilt	1.13	2.01	2.64	3.45	4.01	3.88	3.95	3.74	3.01	2.22	1.36	0.96
21° angle Best winter tilt	1.15	2.00	2.51	3.13	3.56	3.40	3.48	3.35	2.81	2.18	1.37	0.99
53° angle Best summer tilt	1.05	1.92	2.66	3.62	4.37	4.28	4.33	3.98	3.06	2.15	1.27	0.89
Tilt adjusted each month	1.15 21°	2.02 29°	2.64 37°	3.65 45°	4.37 53°	4.57 60°	4.33 53°	4.05 45°	3.01 37°	2.22 29°	1.37 21°	0.99 14°

West Midlands

	Jan	Feb	Mar	Apr	May	Jun	Jul	Aug	Sep	Oct	Nov	Dec
Flat - 90°	0.71	1.35	2.28	3.47	4.51	4.69	4.69	4.04	2.70	1.65	0.90	0.57
Upright - 0°	1.18	1.84	2.18	2.54	2.70	2.59	2.68	2.73	2.41	2.08	1.45	1.03
38° angle Year-round tilt	1.24	2.06	2.74	3.56	4.09	4.05	4.14	3.99	3.14	2.4	1.55	1.06
22° angle Best winter tilt	1.26	2.05	2.60	3.25	3.64	3.55	3.65	3.58	2.95	2.36	1.57	1.09
54° angle Best summer tilt	1.15	1.97	2.75	3.73	4.45	4.46	4.53	4.23	3.19	2.31	1.44	0.97
Tilt adjusted each month	1.26 22°	2.07 30°	2.74 38°	3.76 46°	4.45 54°	4.74 62°	4.53 54°	4.30 46°	3.14 38°	2.40 30°	1.57 22°	1.09 14°

North-East England

	Jan	Feb	Mar	Apr	May	Jun	Jul	Aug	Sep	Oct	Nov	Dec
Flat - 90°	0.61	1.35	2.45	3.88	5.35	5.47	5.13	4.29	2.91	1.66	0.77	0.44
Upright - 0°	1.17	2.03	2.61	3.04	3.31	3.06	3.01	3.05	2.81	2.27	1.38	0.94
35° angle Year-round tilt	1.21	2.23	3.17	4.14	4.9	4.65	4.49	4.32	3.56	2.56	1.44	0.95
19° angle Best winter tilt	1.24	2.22	3.03	3.76	4.33	4.07	3.95	3.85	3.34	2.52	1.46	0.98
51° angle Best summer tilt	1.11	2.12	3.17	4.33	5.33	5.15	4.96	4.59	3.60	2.46	1.34	0.86
Tilt adjusted each month	1.24 19°	2.24 27°	3.17 35°	4.36 43°	5.33 51°	5.53 58°	4.96 51°	4.66 43°	3.56 35°	2.56 27°	1.46 19°	0.98 12°

North-West England

	Jan	Feb	Mar	Apr	May	Jun	Jul	Aug	Sep	Oct	Nov	Dec
Flat - 90°	0.66	1.32	2.30	3.63	4.92	5.00	4.88	3.98	2.73	1.53	0.79	0.50
Upright - 0°	1.16	1.85	2.28	2.74	2.99	2.79	2.82	2.74	2.51	1.95	1.29	0.95
36° angle Year-round tilt	1.21	2.06	2.83	3.79	4.49	4.31	4.31	3.94	3.23	2.23	1.37	0.96
20° angle Best winter tilt	1.24	2.05	2.69	3.46	4.00	3.78	3.8	3.54	3.04	2.19	1.39	0.99
52° angle Best summer tilt	1.12	1.96	2.83	3.97	4.89	4.77	4.73	4.18	3.28	2.16	1.28	0.88
Tilt adjusted each month	1.24 20°	2.07 28°	2.83 36°	3.99 44°	4.89 52°	5.08 60°	4.73 52°	4.25 44°	3.23 36°	2.23 28°	1.39 20°	1.00 12°

Yorkshire and the Humber

	Jan	Feb	Mar	Apr	May	Jun	Jul	Aug	Sep	Oct	Nov	Dec
Flat - 90°	0.62	1.30	2.30	3.51	4.69	4.78	4.64	3.89	2.67	1.54	0.75	0.47
Upright - 0°	1.12	1.86	2.31	2.64	2.86	2.68	2.7	2.68	2.46	1.99	1.25	0.92
36° angle Year-round tilt	1.16	2.06	2.85	3.64	4.24	4.09	4.07	3.83	3.16	2.28	1.32	0.94
20° angle Best winter tilt	1.19	2.04	2.71	3.31	3.77	3.58	3.58	3.42	2.96	2.24	1.34	0.97
52° angle Best summer tilt	1.08	1.96	2.86	3.82	4.63	4.50	4.47	4.08	3.21	2.2	1.24	0.86
Tilt adjusted each month	1.19 20°	2.06 28°	2.85 36°	3.85 44°	4.63 52°	4.83 60°	4.47 52°	4.15 44°	3.16 36°	2.28 28°	1.34 20°	0.97 12°

Central and Southern Scotland

	Jan	Feb	Mar	Apr	May	Jun	Jul	Aug	Sep	Oct	Nov	Dec
Flat - 90°	0.51	1.16	2.03	3.22	4.55	4.66	4.31	3.63	2.42	1.33	0.64	0.38
Upright - 0°	0.99	1.75	2.05	2.46	2.85	2.68	2.58	2.55	2.28	1.76	1.14	0.86
34° angle Year-round tilt	1.02	1.91	2.51	3.32	4.11	3.97	3.77	3.52	2.87	1.99	1.20	0.86
18° angle Best winter tilt	1.04	1.90	2.37	3.00	3.64	3.47	3.31	3.17	2.68	1.95	1.21	0.89
50° angle Best summer tilt	0.94	1.83	2.53	3.49	4.49	4.37	4.15	3.77	2.93	1.93	1.12	0.78
Tilt adjusted each month	1.04 18°	1.92 26°	2.51 34°	3.53 42°	4.49 50°	4.72 58°	4.15 50°	3.86 42°	2.87 34°	1.99 26°	1.21 18°	0.89 10°

North Scotland

	Jan	Feb	Mar	Apr	May	Jun	Jul	Aug	Sep	Oct	Nov	Dec
Flat - 90°	0.47	1.11	2.05	3.30	4.52	4.63	4.31	3.61	2.45	1.31	0.59	0.32
Upright - 0°	0.99	1.74	2.16	2.59	2.88	2.7	2.61	2.57	2.38	1.80	1.12	0.77
33° angle Year-round tilt	1.00	1.88	2.62	3.44	4.08	3.93	3.75	3.50	2.96	2.01	1.16	0.76
17° angle Best winter tilt	1.03	1.87	2.46	3.12	3.62	3.43	3.30	3.15	2.77	1.97	1.18	0.79
49° angle Best summer tilt	0.92	1.80	2.61	3.62	4.46	4.32	4.11	3.75	3.02	1.95	1.08	0.69
Tilt adjusted each month	1.03 17°	1.89 25°	2.62 33°	3.66 41°	4.46 49°	4.69 56°	4.11 49°	3.84 41°	2.96 33°	2.01 25°	1.18 17°	0.79 10°

South Wales

	Jan	Feb	Mar	Apr	May	Jun	Jul	Aug	Sep	Oct	Nov	Dec
Flat - 90°	0.72	1.33	2.21	3.52	4.57	4.75	4.71	3.97	2.71	1.55	0.89	0.59
Upright - 0°	1.14	1.74	2.04	2.54	2.7	2.59	2.66	2.64	2.37	1.87	1.35	0.99
38° angle Year-round tilt	1.21	1.97	2.59	3.61	4.15	4.11	4.16	3.91	3.12	2.18	1.45	1.03
22° angle Best winter tilt	1.23	1.96	2.46	3.29	3.65	3.61	3.68	3.51	2.93	2.14	1.47	1.06
54° angle Best summer tilt	1.13	1.89	2.61	3.78	4.51	4.52	4.56	4.14	3.17	2.11	1.36	0.95
Tilt adjusted each month	1.23 22°	1.98 30°	2.59 38°	3.80 46°	4.51 54°	4.80 62°	4.56 54°	4.20 46°	3.12 38°	2.18 30°	1.47 22°	1.06 14°

North Wales

	Jan	Feb	Mar	Apr	May	Jun	Jul	Aug	Sep	Oct	Nov	Dec
Flat - 90°	0.66	1.32	2.30	3.63	4.92	5.00	4.88	3.98	2.73	1.53	0.79	0.50
Upright - 0°	1.14	1.83	2.26	2.72	2.97	2.77	2.81	2.72	2.49	1.93	1.27	0.92
37° angle Year-round tilt	1.19	2.04	2.81	3.78	4.48	4.30	4.29	3.93	3.22	2.22	1.35	0.95
21° angle Best winter tilt	1.22	2.03	2.68	3.44	3.98	3.77	3.79	3.52	3.02	2.18	1.37	0.98
53° angle Best summer tilt	1.11	1.95	2.82	3.96	4.88	4.76	4.72	4.18	3.27	2.15	1.26	0.87
Tilt adjusted each month	1.22 21°	2.05 29°	2.81 37°	3.98 45°	4.88 53°	5.08 60°	4.72 53°	4.24 45°	3.22 37°	2.22 29°	1.37 21°	0.98 14°

Northern Ireland

	Jan	Feb	Mar	Apr	May	Jun	Jul	Aug	Sep	Oct	Nov	Dec
Flat - 90°	0.61	1.25	2.15	3.39	4.54	4.54	4.30	3.64	2.56	1.42	0.74	0.43
Upright - 0°	1.09	1.79	2.12	2.55	2.79	2.58	2.54	2.51	2.36	1.82	1.25	0.82
35° angle Year-round tilt	1.13	1.98	2.62	3.51	4.12	3.9	3.79	3.56	3.02	2.08	1.32	0.83
19° angle Best winter tilt	1.16	1.97	2.49	3.19	3.66	3.42	3.34	3.19	2.83	2.04	1.34	0.86
51° angle Best summer tilt	1.05	1.89	2.64	3.68	4.49	4.27	4.15	3.79	3.07	2.02	1.24	0.77
Tilt adjusted each month	1.16 19°	1.99 27°	2.62 35°	3.71 43°	4.49 51°	4.59 58°	4.15 51°	3.86 43°	3.02 35°	2.08 27°	1.34 19°	0.86 12°

Solar insolation values – United States of America

Alabama

	Jan	Feb	Mar	Apr	May	Jun	Jul	Aug	Sep	Oct	Nov	Dec
Flat - 90°	2.32	2.9	4.06	5.01	5.54	5.64	5.7	5.14	4.58	3.84	2.68	2.17
Upright - 0°	2.94	2.85	3.06	2.67	2.30	2.15	2.22	2.46	3.03	3.72	3.29	3.00
58° angle Year-round tilt	3.29	3.64	4.60	5.04	5.10	5.02	5.14	4.96	4.98	4.86	3.76	3.22
42° angle Best winter tilt	3.45	3.68	4.46	4.65	4.52	4.36	4.49	4.50	4.75	4.90	3.92	3.42
74° angle Best summer tilt	2.96	3.42	4.5	5.18	5.46	5.45	5.55	5.20	4.96	4.56	3.39	2.85
Tilt adjusted each month	3.45 42°	3.69 50°	4.60 58°	5.18 66°	5.46 74°	5.64 82°	5.55 74°	5.22 66°	4.98 58°	4.92 50°	3.92 42°	3.45 34°

Alaska

	Jan	Feb	Mar	Apr	May	Jun	Jul	Aug	Sep	Oct	Nov	Dec
Flat - 90°	0.40	1.07	2.22	3.73	4.95	5.26	4.86	3.86	2.46	1.37	0.58	0.00
Upright - 0°	0.88	1.75	2.51	3.09	3.22	3.08	2.98	2.83	2.47	2.00	1.20	0.00
32° angle Year-round tilt	0.89	1.89	2.97	4.06	4.50	4.47	4.24	3.81	3.03	2.21	1.23	0.00
16° angle Best winter tilt	0.92	1.88	2.83	3.68	3.99	3.90	3.73	3.43	2.84	2.17	1.25	0.00
48° angle Best summer tilt	0.82	1.80	2.97	4.25	4.93	4.93	4.66	4.07	3.09	2.13	1.14	0.00
Tilt adjusted each month	0.92 16°	1.90 24°	2.97 32°	4.27 40°	4.93 48°	5.35 56°	4.66 48°	4.16 40°	3.03 32°	2.21 24°	1.25 16°	0.00 8°

Arizona

	Jan	Feb	Mar	Apr	May	Jun	Jul	Aug	Sep	Oct	Nov	Dec
Flat - 90°	3.20	4.07	5.45	6.62	7.37	7.52	6.78	5.97	5.45	4.48	3.50	2.95
Upright - 0°	4.63	4.59	4.29	3.33	2.57	2.28	2.34	2.70	3.60	4.51	4.76	4.63
57° angle Year-round tilt	4.92	5.55	6.45	6.83	6.74	6.57	6.08	5.78	6.03	5.82	5.24	4.74
41° angle Best winter tilt	5.28	5.74	6.34	6.33	5.87	5.58	5.26	5.22	5.78	5.93	5.57	5.14
73° angle Best summer tilt	4.28	5.04	6.20	6.96	7.26	7.23	6.60	6.05	5.95	5.39	4.60	4.06
Tilt adjusted each month	5.28 41°	5.74 49°	6.45 57°	6.97 65°	7.26 73°	7.52 80°	6.60 73°	6.07 65°	6.03 57°	5.94 49°	5.57 41°	5.24 34°

Arkansas

	Jan	Feb	Mar	Apr	May	Jun	Jul	Aug	Sep	Oct	Nov	Dec
Flat - 90°	2.39	3.03	4.01	5.19	5.51	5.91	6.02	5.52	4.74	3.71	2.60	2.15
Upright - 0°	3.14	3.09	3.07	2.80	2.33	2.22	2.31	2.64	3.20	3.62	3.23	3.02
55° angle Year-round tilt	3.45	3.88	4.56	5.24	5.10	5.26	5.44	5.36	5.20	4.70	3.66	3.21
39° angle Best winter tilt	3.64	3.94	4.44	4.84	4.52	4.57	4.76	4.86	4.98	4.74	3.82	3.42
71° angle Best summer tilt	3.08	3.62	4.46	5.38	5.44	5.72	5.88	5.61	5.16	4.41	3.30	2.83
Tilt adjusted each month	3.64 39°	3.94 47°	4.56 55°	5.38 63°	5.44 71°	5.91 78°	5.88 71°	5.63 63°	5.20 55°	4.75 47°	3.82 39°	3.45 32°

California

	Jan	Feb	Mar	Apr	May	Jun	Jul	Aug	Sep	Oct	Nov	Dec
Flat - 90°	2.18	3.09	4.65	6.08	7.21	7.93	7.79	7.02	5.64	4.12	2.63	1.99
Upright - 0°	3.35	3.68	4.09	3.53	2.96	2.69	2.82	3.46	4.34	4.83	3.96	3.37
51° angle Year-round tilt	3.52	4.36	5.72	6.32	6.64	6.91	6.95	6.95	6.64	5.86	4.23	3.41
35° angle Best winter tilt	3.73	4.46	5.61	5.83	5.79	5.84	5.95	6.25	6.39	5.98	4.47	3.66
67° angle Best summer tilt	3.12	4.02	5.52	6.49	7.16	7.63	7.60	7.29	6.52	5.40	3.75	2.97
Tilt adjusted each month	3.73 35°	4.46 43°	5.72 51°	6.49 59°	7.16 67°	7.95 74°	7.60 67°	7.31 59°	6.64 51°	5.98 43°	4.47 35°	3.71 28°

Colorado

	Jan	Feb	Mar	Apr	May	Jun	Jul	Aug	Sep	Oct	Nov	Dec
Flat - 90°	2.41	3.27	4.49	5.42	6.28	6.70	6.35	5.68	5.03	3.90	2.67	2.18
Upright - 0°	4.20	4.18	4.05	3.25	2.80	2.60	2.64	2.99	3.90	4.66	4.30	4.21
50° angle Year-round tilt	4.27	4.85	5.58	5.62	5.81	5.90	5.72	5.60	5.87	5.60	4.52	4.12
34° angle Best winter tilt	4.58	4.99	5.47	5.17	5.10	5.05	4.96	5.05	5.63	5.71	4.79	4.47
66° angle Best summer tilt	3.72	4.43	5.39	5.78	6.24	6.47	6.21	5.87	5.79	5.17	3.98	3.53
Tilt adjusted each month	4.58 34°	4.99 42°	5.58 50°	5.78 58°	6.24 66°	6.73 74°	6.21 66°	5.89 58°	5.87 50°	5.71 42°	4.79 34°	4.56 26°

Connecticut

	Jan	Feb	Mar	Apr	May	Jun	Jul	Aug	Sep	Oct	Nov	Dec
Flat - 90°	1.88	2.73	3.76	4.52	5.23	5.80	5.54	5.00	4.11	3.01	1.97	1.63
Upright - 0°	3.20	3.50	3.41	2.83	2.57	2.52	2.54	2.79	3.23	3.45	3.00	3.01
48° angle Year-round tilt	3.29	4.05	4.62	4.65	4.85	5.07	5.01	4.92	4.73	4.17	3.21	3.00
32° angle Best winter tilt	3.49	4.14	4.50	4.28	4.28	4.38	4.37	4.45	4.51	4.20	3.35	3.21
64° angle Best summer tilt	2.91	3.73	4.50	4.80	5.20	5.53	5.42	5.16	4.71	3.92	2.88	2.61
Tilt adjusted each month	3.49 32°	4.14 40°	4.62 48°	4.80 56°	5.20 64°	5.81 72°	5.42 64°	5.19 56°	4.73 48°	4.21 40°	3.35 32°	3.25 24°

Delaware

	Jan	Feb	Mar	Apr	May	Jun	Jul	Aug	Sep	Oct	Nov	Dec
Flat - 90°	1.97	2.73	3.69	4.65	5.27	5.67	5.50	5.05	4.25	3.33	2.24	1.77
Upright - 0°	2.99	3.16	3.10	2.77	2.46	2.36	2.40	2.68	3.15	3.65	3.21	2.93
51° angle Year-round tilt	3.17	3.78	4.35	4.73	4.86	5.01	4.95	4.92	4.78	4.51	3.50	3.00
35° angle Best winter tilt	3.34	3.84	4.22	4.36	4.29	4.34	4.32	4.44	4.56	4.54	3.66	3.20
67° angle Best summer tilt	2.82	3.52	4.26	4.89	5.21	5.47	5.36	5.17	4.77	4.23	3.14	2.64
Tilt adjusted each month	3.34 35°	3.85 43°	4.35 51°	4.89 59°	5.21 67°	5.68 74°	5.36 67°	5.20 59°	4.78 51°	4.56 43°	3.66 35°	3.23 28°

Florida

	Jan	Feb	Mar	Apr	May	Jun	Jul	Aug	Sep	Oct	Nov	Dec
Flat - 90°	3.56	4.44	5.29	6.11	6.49	5.97	5.55	5.26	4.76	4.52	3.78	3.35
Upright - 0°	4.03	4.07	3.44	2.62	2.06	1.88	1.91	2.15	2.68	3.68	4.04	4.07
63° angle Year-round tilt	4.66	5.39	5.80	6.11	5.98	5.38	5.08	5.06	4.96	5.27	4.85	4.51
47° angle Best winter tilt	4.96	5.56	5.69	5.68	5.28	4.69	4.50	4.62	4.75	5.33	5.11	4.85
79° angle Best summer tilt	4.11	4.94	5.61	6.23	6.39	5.82	5.45	5.26	4.93	4.93	4.32	3.92
Tilt adjusted each month	4.96 47°	5.56 55°	5.80 63°	6.23 71°	6.39 79°	5.97 86°	5.45 79°	5.28 71°	4.96 63°	5.34 55°	5.11 47°	4.92 40°

Georgia

	Jan	Feb	Mar	Apr	May	Jun	Jul	Aug	Sep	Oct	Nov	Dec
Flat - 90°	2.50	3.12	4.28	5.19	5.80	5.59	5.72	5.17	4.46	3.91	2.82	2.36
Upright - 0°	3.21	3.14	3.23	2.74	2.34	2.13	2.21	2.46	2.92	3.76	3.48	3.32
56° angle Year-round tilt	3.55	3.96	4.86	5.24	5.36	5.00	5.18	5.01	4.82	4.92	3.96	3.52
40° angle Best winter tilt	3.74	4.02	4.74	4.86	4.74	4.37	4.55	4.55	4.60	4.97	4.15	3.76
72° angle Best summer tilt	3.17	3.69	4.74	5.38	5.72	5.42	5.58	5.23	4.80	4.60	3.55	3.09
Tilt adjusted each month	3.74 40°	4.03 48°	4.86 56°	5.38 64°	5.72 72°	5.59 80°	5.58 72°	5.25 64°	4.82 56°	4.98 48°	4.15 40°	3.80 32°

Hawaii

	Jan	Feb	Mar	Apr	May	Jun	Jul	Aug	Sep	Oct	Nov	Dec
Flat - 90°	4.09	5.06	5.85	6.58	6.97	7.35	7.21	6.88	6.36	5.41	4.34	3.90
Upright - 0°	4.06	4.12	3.33	2.33	1.77	2.80	1.69	2.01	3.00	3.96	4.22	4.12
69° angle Year-round tilt	4.98	5.84	6.24	6.50	6.43	7.37	6.52	6.63	6.60	6.10	5.24	4.83
53° angle Best winter tilt	5.29	6.03	6.13	6.04	5.65	6.92	5.65	6.03	6.34	6.20	5.56	5.19
85° angle Best summer tilt	4.40	5.34	6.03	6.62	6.89	7.43	7.08	6.88	6.50	5.67	4.65	4.21
Tilt adjusted each month	5.29 53°	6.03 61°	6.24 69°	6.63 77°	6.89 85°	7.46 92°	7.08 85°	6.89 77°	6.60 69°	6.21 61°	5.56 53°	5.27 46°

Idaho

	Jan	Feb	Mar	Apr	May	Jun	Jul	Aug	Sep	Oct	Nov	Dec
Flat - 90°	1.72	2.66	4.07	5.38	6.36	7.24	7.37	6.44	4.99	3.43	1.97	1.49
Upright - 0°	3.17	3.67	4.01	3.53	3.08	2.98	3.18	3.66	4.33	4.56	3.35	3.00
47° angle Year-round tilt	3.21	4.15	5.28	5.72	5.92	6.26	6.56	6.51	6.14	5.28	3.49	2.96
31° angle Best winter tilt	3.41	4.25	5.16	5.26	5.17	5.31	5.61	5.85	5.89	5.38	3.67	3.17
63° angle Best summer tilt	2.83	3.82	5.10	5.88	6.38	6.89	7.16	6.83	6.04	4.88	3.11	2.57
Tilt adjusted each month	3.41 31°	4.25 39°	5.28 47°	5.88 55°	6.38 63°	7.26 70°	7.16 63°	6.85 55°	6.14 47°	5.38 39°	3.67 31°	3.21 24°

Illinois

	Jan	Feb	Mar	Apr	May	Jun	Jul	Aug	Sep	Oct	Nov	Dec
Flat - 90°	1.92	2.55	3.62	4.69	5.40	5.90	6.12	5.38	4.57	3.32	2.10	1.71
Upright - 0°	2.91	2.91	3.06	2.83	2.54	2.44	2.59	2.86	3.48	3.70	2.95	2.84
50° angle Year-round tilt	3.08	3.48	4.28	4.80	5.01	5.23	5.52	5.29	5.24	4.53	3.22	2.90
34° angle Best winter tilt	3.24	3.53	4.15	4.42	4.42	4.52	4.80	4.78	5.01	4.58	3.36	3.09
66° angle Best summer tilt	2.74	3.25	4.19	4.95	5.36	5.71	5.99	5.54	5.19	4.24	2.90	2.55
Tilt adjusted each month	3.24 34°	3.54 42°	4.28 50°	4.95 58°	5.36 66°	5.92 74°	5.99 66°	5.57 58°	5.24 50°	4.59 42°	3.36 34°	3.12 26°

Indiana

	Jan	Feb	Mar	Apr	May	Jun	Jul	Aug	Sep	Oct	Nov	Dec
Flat - 90°	1.83	2.54	3.55	4.49	5.19	5.91	5.98	5.24	4.55	3.26	2.02	1.57
Upright - 0°	2.70	2.89	2.99	2.70	2.46	2.44	2.56	2.80	3.46	3.60	2.78	2.47
50° angle Year-round tilt	2.87	3.46	4.18	4.57	4.80	5.24	5.40	5.15	5.21	4.42	3.05	2.56
34° angle Best winter tilt	3.02	3.51	4.05	4.21	4.25	4.53	4.70	4.65	4.98	4.46	3.17	2.72
66° angle Best summer tilt	2.57	3.24	4.10	4.71	5.14	5.72	5.85	5.39	5.17	4.15	2.76	2.27
Tilt adjusted each month	3.02 34°	3.52 42°	4.18 50°	4.72 58°	5.14 66°	5.93 74°	5.85 66°	5.42 58°	5.21 50°	4.48 42°	3.17 34°	2.74 26°

Iowa

	Jan	Feb	Mar	Apr	May	Jun	Jul	Aug	Sep	Oct	Nov	Dec
Flat - 90°	1.91	2.56	3.62	4.56	5.39	6.12	6.02	5.32	4.41	3.14	2.02	1.64
Upright - 0°	3.26	3.15	3.23	2.84	2.62	2.59	2.67	2.94	3.50	3.67	3.10	3.04
48° angle Year-round tilt	3.35	3.68	4.40	4.68	4.99	5.33	5.43	5.26	5.13	4.42	3.31	3.04
32° angle Best winter tilt	3.55	3.74	4.27	4.31	4.40	4.59	4.71	4.74	4.90	4.46	3.46	3.25
64° angle Best summer tilt	2.96	3.42	4.30	4.84	5.36	5.83	5.90	5.52	5.09	4.14	2.97	2.65
Tilt adjusted each month	3.55 32°	3.75 40°	4.40 48°	4.84 56°	5.36 64°	6.13 72°	5.90 64°	5.55 56°	5.13 48°	4.47 40°	3.46 32°	3.29 24°

Kansas

	Jan	Feb	Mar	Apr	May	Jun	Jul	Aug	Sep	Oct	Nov	Dec
Flat - 90°	2.15	2.72	3.91	4.80	5.61	6.15	6.37	5.48	4.67	3.41	2.30	1.92
Upright - 0°	3.40	3.13	3.32	2.84	2.56	2.46	2.60	2.86	3.50	3.76	3.32	3.32
51° angle Year-round tilt	3.57	3.75	4.66	4.89	5.16	5.41	5.70	5.36	5.33	4.64	3.61	3.36
35° angle Best winter tilt	3.78	3.81	4.53	4.50	4.54	4.65	4.93	4.83	5.09	4.68	3.78	3.60
67° angle Best summer tilt	3.16	3.50	4.55	5.05	5.54	5.93	6.21	5.63	5.29	4.35	3.24	2.93
Tilt adjusted each month	3.78 35°	3.81 43°	4.66 51°	5.06 59°	5.54 67°	6.17 74°	6.21 67°	5.66 59°	5.33 51°	4.69 43°	3.78 35°	3.64 28°

Louisiana

	Jan	Feb	Mar	Apr	May	Jun	Jul	Aug	Sep	Oct	Nov	Dec
Flat - 90°	2.59	3.25	4.31	5.21	5.79	5.65	5.66	5.30	4.72	4.03	3.04	2.56
Upright - 0°	2.95	3.02	3.01	2.57	2.19	2.00	2.07	2.34	2.89	3.55	3.42	3.22
60° angle Year-round tilt	3.41	3.95	4.76	5.23	5.35	5.07	5.14	5.11	5.02	4.85	4.03	3.54
44° angle Best winter tilt	3.57	4.00	4.63	4.86	4.75	4.43	4.53	4.65	4.80	4.89	4.21	3.77
76° angle Best summer tilt	3.08	3.69	4.65	5.35	5.71	5.48	5.53	5.33	4.99	4.55	3.63	3.13
Tilt adjusted each month	3.57 44°	4.01 52°	4.76 60°	5.35 68°	5.71 76°	5.65 84°	5.53 76°	5.35 68°	5.02 60°	4.90 52°	4.21 44°	3.80 36°

Maine

	Jan	Feb	Mar	Apr	May	Jun	Jul	Aug	Sep	Oct	Nov	Dec
Flat - 90°	1.69	2.57	3.64	4.43	4.87	5.35	5.28	4.85	3.78	2.56	1.60	1.33
Upright - 0°	3.25	3.61	3.52	2.91	2.52	2.50	2.56	2.84	3.10	3.01	2.51	2.57
46° angle Year-round tilt	3.28	4.07	4.63	4.59	4.50	4.64	4.69	4.80	4.39	3.60	2.68	2.58
30° angle Best winter tilt	3.48	4.16	4.50	4.21	3.96	4.01	4.09	4.32	4.17	3.60	2.78	2.74
62° angle Best summer tilt	2.88	3.75	4.52	4.76	4.84	5.09	5.10	5.05	4.39	3.41	2.43	2.27
Tilt adjusted each month	3.48 30°	4.16 38°	4.63 46°	4.76 54°	4.84 62°	5.37 70°	5.10 62°	5.09 54°	4.39 46°	3.62 38°	2.78 30°	2.76 22°

Maryland

	Jan	Feb	Mar	Apr	May	Jun	Jul	Aug	Sep	Oct	Nov	Dec
Flat - 90°	1.94	2.68	3.67	4.63	5.31	5.58	5.53	4.94	4.19	3.33	2.19	1.71
Upright - 0°	2.89	3.06	3.06	2.75	2.47	2.33	2.40	2.62	3.09	3.63	3.08	2.75
51° angle Year-round tilt	3.09	3.68	4.31	4.70	4.89	4.93	4.97	4.80	4.70	4.50	3.37	2.84
35° angle Best winter tilt	3.24	3.73	4.18	4.33	4.31	4.27	4.34	4.34	4.47	4.53	3.52	3.02
67° angle Best summer tilt	2.76	3.44	4.23	4.86	5.24	5.38	5.39	5.04	4.69	4.22	3.04	2.51
Tilt adjusted each month	3.24 35°	3.74 43°	4.31 51°	4.87 59°	5.24 67°	5.59 74°	5.39 67°	5.08 59°	4.70 51°	4.55 43°	3.52 35°	3.04 28°

Massachusetts

	Jan	Feb	Mar	Apr	May	Jun	Jul	Aug	Sep	Oct	Nov	Dec
Flat - 90°	1.80	2.62	3.61	4.46	5.13	5.50	5.60	4.99	4.04	2.93	1.89	1.54
Upright - 0°	3.15	3.40	3.30	2.83	2.56	2.47	2.59	2.81	3.22	3.41	2.94	2.91
48° angle Year-round tilt	3.22	3.91	4.44	4.59	4.75	4.79	5.05	4.91	4.67	4.10	3.13	2.90
32° angle Best winter tilt	3.42	3.99	4.32	4.22	4.19	4.15	4.40	4.43	4.45	4.12	3.27	3.10
64° angle Best summer tilt	2.85	3.62	4.34	4.74	5.10	5.24	5.49	5.16	4.66	3.86	2.82	2.53
Tilt adjusted each month	3.42 32°	3.99 40°	4.44 48°	4.75 56°	5.10 64°	5.51 72°	5.49 64°	5.19 56°	4.67 48°	4.14 40°	3.27 32°	3.14 24°

Michigan

	Jan	Feb	Mar	Apr	May	Jun	Jul	Aug	Sep	Oct	Nov	Dec
Flat - 90°	1.78	2.51	3.44	4.45	5.24	6.07	5.93	5.03	4.13	2.79	1.69	1.41
Upright - 0°	3.13	3.22	3.11	2.84	2.61	2.64	2.70	2.86	3.33	3.20	2.47	2.55
47° angle Year-round tilt	3.20	3.71	4.20	4.59	4.86	5.29	5.36	4.97	4.81	3.86	2.67	2.56
31° angle Best winter tilt	3.39	3.78	4.07	4.22	4.29	4.55	4.65	4.49	4.59	3.87	2.77	2.73
63° angle Best summer tilt	2.83	3.44	4.11	4.74	5.21	5.78	5.82	5.22	4.78	3.64	2.43	2.25
Tilt adjusted each month	3.39 31°	3.78 39°	4.20 47°	4.74 55°	5.21 63°	6.09 70°	5.82 63°	5.25 55°	4.81 47°	3.89 39°	2.77 31°	2.75 24°

Minnesota

	Jan	Feb	Mar	Apr	May	Jun	Jul	Aug	Sep	Oct	Nov	Dec
Flat - 90°	1.71	2.55	3.44	4.54	5.32	5.96	6.05	5.19	4.00	2.78	1.76	1.37
Upright - 0°	3.41	3.65	3.33	3.03	2.75	2.73	2.87	3.08	3.39	3.49	3.03	2.81
45° angle Year-round tilt	3.39	4.08	4.36	4.76	4.96	5.19	5.40	5.20	4.76	4.08	3.15	2.77
29° angle Best winter tilt	3.62	4.18	4.24	4.37	4.36	4.47	4.68	4.68	4.54	4.12	3.30	2.97
61° angle Best summer tilt	2.97	3.74	4.25	4.91	5.33	5.68	5.87	5.46	4.73	3.83	2.82	2.42
Tilt adjusted each month	3.62 29°	4.18 37°	4.36 45°	4.92 53°	5.33 61°	5.99 68°	5.87 61°	5.48 53°	4.76 45°	4.13 37°	3.30 29°	3.00 22°

Mississippi

	Jan	Feb	Mar	Apr	May	Jun	Jul	Aug	Sep	Oct	Nov	Dec
Flat - 90°	2.52	3.24	4.24	5.34	5.89	5.98	5.80	5.36	4.75	3.99	2.94	2.39
Upright - 0°	3.08	3.20	3.09	2.72	2.30	2.12	2.17	2.46	3.04	3.71	3.53	3.17
58° angle Year-round tilt	3.48	4.07	4.75	5.39	5.43	5.33	5.25	5.17	5.12	4.94	4.05	3.44
42° angle Best winter tilt	3.66	4.14	4.62	5.00	4.80	4.62	4.60	4.70	4.90	4.99	4.24	3.65
74° angle Best summer tilt	3.12	3.79	4.64	5.52	5.80	5.79	5.66	5.41	5.09	4.63	3.64	3.03
Tilt adjusted each month	3.66 42°	4.14 50°	4.75 58°	5.52 66°	5.80 74°	5.98 82°	5.66 74°	5.43 66°	5.12 58°	5.00 50°	4.24 42°	3.69 34°

Missouri

	Jan	Feb	Mar	Apr	May	Jun	Jul	Aug	Sep	Oct	Nov	Dec
Flat - 90°	2.09	2.71	3.85	4.94	5.53	5.95	6.18	5.52	4.62	3.42	2.23	1.89
Upright - 0°	3.13	3.04	3.21	2.90	2.52	2.40	2.54	2.85	3.42	3.70	3.07	3.10
51° angle Year-round tilt	3.32	3.67	4.54	5.05	5.11	5.27	5.57	5.41	5.24	4.59	3.37	3.17
35° angle Best winter tilt	3.51	3.72	4.41	4.65	4.51	4.55	4.84	4.89	5.01	4.63	3.52	3.38
67° angle Best summer tilt	2.95	3.42	4.43	5.20	5.47	5.74	6.04	5.67	5.20	4.30	3.04	2.77
Tilt adjusted each month	3.51 35°	3.73 43°	4.54 51°	5.20 59°	5.47 67°	5.96 74°	6.04 67°	5.70 59°	5.24 51°	4.65 43°	3.52 35°	3.42 28°

Montana

	Jan	Feb	Mar	Apr	May	Jun	Jul	Aug	Sep	Oct	Nov	Dec
Flat - 90°	1.64	2.51	3.65	4.72	5.60	6.31	6.58	5.70	4.30	2.83	1.83	1.37
Upright - 0°	2.59	3.19	3.41	3.22	2.95	2.94	3.18	3.45	3.62	3.31	2.73	2.25
43° angle Year-round tilt	2.73	3.67	4.48	4.98	5.21	5.48	5.87	5.71	5.04	3.95	2.94	2.34
27° angle Best winter tilt	2.83	3.69	4.33	4.58	4.59	4.71	5.08	5.14	4.79	3.94	3.03	2.44
59° angle Best summer tilt	2.48	3.45	4.41	5.14	5.61	6.00	6.39	6.00	5.04	3.76	2.70	2.12
Tilt adjusted each month	2.83 27°	3.71 35°	4.48 43°	5.15 51°	5.61 59°	6.34 66°	6.39 59°	6.04 51°	5.04 43°	3.97 35°	3.03 27°	2.45 20°

Nebraska

	Jan	Feb	Mar	Apr	May	Jun	Jul	Aug	Sep	Oct	Nov	Dec
Flat - 90°	2.09	2.68	3.83	4.78	5.57	6.35	6.35	5.48	4.63	3.35	2.26	1.86
Upright - 0°	3.55	3.26	3.39	2.93	2.64	2.60	2.71	2.97	3.63	3.90	3.51	3.50
49° angle Year-round tilt	3.65	3.82	4.65	4.91	5.16	5.54	5.73	5.41	5.38	4.70	3.72	3.47
33° angle Best winter tilt	3.88	3.90	4.53	4.53	4.55	4.77	4.96	4.89	5.15	4.76	3.92	3.74
65° angle Best summer tilt	3.20	3.55	4.53	5.07	5.54	6.06	6.22	5.67	5.33	4.39	3.32	3.00
Tilt adjusted each month	3.88 33°	3.90 41°	4.65 49°	5.07 57°	5.54 65°	6.36 72°	6.22 65°	5.70 57°	5.38 49°	4.77 41°	3.92 33°	3.79 26°

Nevada

	Jan	Feb	Mar	Apr	May	Jun	Jul	Aug	Sep	Oct	Nov	Dec
Flat - 90°	2.29	3.15	4.61	5.85	7.01	7.77	7.76	6.89	5.60	4.06	2.60	2.11
Upright - 0°	3.84	3.94	4.16	3.48	2.98	2.74	2.90	3.49	4.42	4.91	4.10	3.95
51° angle Year-round tilt	3.95	4.60	5.75	6.10	6.47	6.77	6.93	6.85	6.67	5.89	4.33	3.91
35° angle Best winter tilt	4.21	4.72	5.64	5.61	5.64	5.72	5.93	6.15	6.41	6.01	4.58	4.22
67° angle Best summer tilt	3.46	4.22	5.55	6.26	6.97	7.47	7.58	7.17	6.55	5.42	3.83	3.36
Tilt adjusted each month	4.21 35°	4.72 43°	5.75 51°	6.26 59°	6.97 67°	7.79 74°	7.58 67°	7.20 59°	6.67 51°	6.01 43°	4.58 35°	4.30 28°

New Hampshire

	Jan	Feb	Mar	Apr	May	Jun	Jul	Aug	Sep	Oct	Nov	Dec
Flat - 90°	1.74	2.62	3.61	4.43	4.92	5.42	5.47	4.91	3.92	2.72	1.71	1.46
Upright - 0°	3.17	3.54	3.38	2.85	2.51	2.48	2.58	2.82	3.16	3.17	2.63	2.86
47° angle Year-round tilt	3.23	4.03	4.51	4.57	4.55	4.71	4.90	4.85	4.54	3.80	2.81	2.84
31° angle Best winter tilt	3.42	4.12	4.38	4.20	4.01	4.08	4.26	4.36	4.31	3.81	2.92	3.03
63° angle Best summer tilt	2.85	3.72	4.40	4.73	4.89	5.16	5.32	5.10	4.53	3.59	2.54	2.47
Tilt adjusted each month	3.42 31°	4.12 39°	4.51 47°	4.74 55°	4.89 63°	5.43 70°	5.32 63°	5.13 55°	4.54 47°	3.83 39°	2.92 31°	3.07 24°

New Jersey

	Jan	Feb	Mar	Apr	May	Jun	Jul	Aug	Sep	Oct	Nov	Dec
Flat - 90°	1.92	2.71	3.69	4.57	5.21	5.61	5.51	4.97	4.15	3.13	2.07	1.66
Upright - 0°	3.03	3.25	3.18	2.77	2.49	2.39	2.45	2.69	3.15	3.47	2.99	2.83
50° angle Year-round tilt	3.18	3.84	4.41	4.66	4.81	4.96	4.97	4.85	4.70	4.26	3.24	2.88
34° angle Best winter tilt	3.35	3.91	4.28	4.28	4.25	4.29	4.33	4.38	4.48	4.28	3.38	3.07
66° angle Best summer tilt	2.82	3.57	4.32	4.81	5.16	5.42	5.38	5.10	4.69	4.01	2.93	2.53
Tilt adjusted each month	3.35 34°	3.91 42°	4.41 50°	4.82 58°	5.16 66°	5.63 74°	5.38 66°	5.13 58°	4.70 50°	4.30 42°	3.38 34°	3.10 26°

New Mexico

	Jan	Feb	Mar	Apr	May	Jun	Jul	Aug	Sep	Oct	Nov	Dec
Flat - 90°	3.04	3.85	5.14	6.32	7.04	7.22	6.51	5.81	5.41	4.39	3.29	2.81
Upright - 0°	4.80	4.49	4.27	3.39	2.70	2.42	2.44	2.79	3.80	4.73	4.83	4.87
54° angle Year-round tilt	4.98	5.39	6.20	6.48	6.47	6.33	5.85	5.66	6.11	5.93	5.18	4.85
38° angle Best winter tilt	5.36	5.56	6.10	5.98	5.66	5.40	5.08	5.12	5.87	6.05	5.52	5.28
70° angle Best summer tilt	4.31	4.91	5.96	6.65	6.96	6.95	6.34	5.92	6.02	5.47	4.54	4.13
Tilt adjusted each month	5.36 38°	5.56 46°	6.20 54°	6.65 62°	6.96 70°	7.22 78°	6.34 70°	5.95 62°	6.11 54°	6.05 46°	5.52 38°	5.40 30°

New York

	Jan	Feb	Mar	Apr	May	Jun	Jul	Aug	Sep	Oct	Nov	Dec
Flat - 90°	1.74	2.60	3.57	4.34	5.04	5.64	5.55	4.97	3.95	2.80	1.76	1.47
Upright - 0°	3.02	3.39	3.27	2.76	2.54	2.52	2.58	2.82	3.15	3.22	2.63	2.73
47° angle Year-round tilt	3.09	3.89	4.40	4.46	4.67	4.92	5.02	4.91	4.56	3.88	2.83	2.73
31° angle Best winter tilt	3.28	3.97	4.27	4.10	4.13	4.26	4.37	4.43	4.35	3.89	2.94	2.91
63° angle Best summer tilt	2.74	3.60	4.29	4.61	5.01	5.37	5.44	5.16	4.55	3.66	2.56	2.39
Tilt adjusted each month	3.28 31°	3.97 39°	4.40 47°	4.62 55°	5.01 63°	5.65 70°	5.44 63°	5.19 55°	4.56 47°	3.91 39°	2.94 31°	2.94 24°

North Carolina

	Jan	Feb	Mar	Apr	May	Jun	Jul	Aug	Sep	Oct	Nov	Dec
Flat - 90°	2.43	3.05	4.20	5.22	5.73	5.80	5.68	5.08	4.43	3.76	2.73	2.30
Upright - 0°	3.39	3.24	3.33	2.87	2.43	2.25	2.30	2.52	3.05	3.82	3.64	3.56
54° angle Year-round tilt	3.66	4.00	4.87	5.29	5.29	5.16	5.15	4.93	4.86	4.87	4.03	3.67
38° angle Best winter tilt	3.87	4.07	4.75	4.89	4.69	4.49	4.51	4.47	4.64	4.93	4.24	3.94
70° angle Best summer tilt	3.24	3.72	4.74	5.44	5.66	5.61	5.55	5.15	4.84	4.55	3.60	3.20
Tilt adjusted each month	3.87 38°	4.07 46°	4.87 54°	5.44 62°	5.66 70°	5.80 78°	5.55 70°	5.18 62°	4.86 54°	4.94 46°	4.24 38°	4 30°

North Dakota

	Jan	Feb	Mar	Apr	May	Jun	Jul	Aug	Sep	Oct	Nov	Dec
Flat - 90°	1.48	2.35	3.47	4.77	5.72	6.26	6.49	5.48	4.06	2.73	1.71	1.27
Upright - 0°	2.36	3.02	3.29	3.29	3.06	2.98	3.22	3.39	3.47	3.25	2.61	2.15
43° angle Year-round tilt	2.49	3.46	4.28	5.01	5.27	5.43	5.79	5.50	4.78	3.85	2.80	2.22
27° angle Best winter tilt	2.58	3.48	4.13	4.60	4.63	4.67	5.01	4.95	4.54	3.83	2.88	2.32
59° angle Best summer tilt	2.28	3.26	4.22	5.19	5.68	5.96	6.30	5.78	4.78	3.67	2.57	2.01
Tilt adjusted each month	2.58 27°	3.50 35°	4.28 43°	5.19 51°	5.68 59°	6.30 66°	6.30 59°	5.82 51°	4.78 43°	3.87 35°	2.88 27°	2.32 20°

Ohio

	Jan	Feb	Mar	Apr	May	Jun	Jul	Aug	Sep	Oct	Nov	Dec
Flat - 90°	1.77	2.49	3.31	4.47	5.18	5.64	5.76	5.11	4.26	3.08	1.86	1.46
Upright - 0°	2.64	2.86	2.76	2.70	2.47	2.39	2.51	2.74	3.23	3.37	2.51	2.28
50° angle Year-round tilt	2.81	3.43	3.86	4.54	4.77	4.98	5.18	4.99	4.84	4.16	2.78	2.38
34° angle Best winter tilt	2.95	3.47	3.73	4.17	4.21	4.31	4.50	4.50	4.61	4.18	2.88	2.51
66° angle Best summer tilt	2.53	3.21	3.80	4.69	5.12	5.44	5.62	5.24	4.82	3.92	2.54	2.12
Tilt adjusted each month	2.95 34°	3.47 42°	3.86 50°	4.70 58°	5.12 66°	5.66 74°	5.62 66°	5.28 58°	4.84 50°	4.20 42°	2.88 34°	2.52 26°

Oklahoma

	Jan	Feb	Mar	Apr	May	Jun	Jul	Aug	Sep	Oct	Nov	Dec
Flat - 90°	2.59	3.20	4.29	5.40	5.84	6.32	6.73	5.86	4.81	3.72	2.75	2.33
Upright - 0°	3.70	3.43	3.39	2.94	2.44	2.31	2.46	2.79	3.31	3.73	3.64	3.59
54° angle Year-round tilt	3.97	4.23	4.98	5.47	5.38	5.59	6.03	5.70	5.32	4.79	4.03	3.71
38° angle Best winter tilt	4.21	4.31	4.86	5.05	4.75	4.82	5.22	5.15	5.09	4.83	4.25	3.98
70° angle Best summer tilt	3.50	3.92	4.85	5.63	5.76	6.10	6.55	5.97	5.28	4.49	3.62	3.23
Tilt adjusted each month	4.21 38°	4.31 46°	4.98 54°	5.63 62°	5.76 70°	6.33 78°	6.55 70°	5.99 62°	5.32 54°	4.85 46°	4.25 38°	4.04 30°

Oregon

	Jan	Feb	Mar	Apr	May	Jun	Jul	Aug	Sep	Oct	Nov	Dec
Flat - 90°	1.41	2.24	3.26	4.33	5.26	6.19	6.85	6.04	4.55	2.82	1.49	1.16
Upright - 0°	2.40	2.91	3.03	2.83	2.68	2.74	3.06	3.48	3.91	3.47	2.23	2.04
46° angle Year-round tilt	2.49	3.35	4.02	4.47	4.86	5.35	6.07	6.07	5.53	4.10	2.41	2.09
30° angle Best winter tilt	2.61	3.39	3.89	4.10	4.27	4.58	5.20	5.44	5.28	4.12	2.49	2.20
62° angle Best summer tilt	2.23	3.12	3.95	4.64	5.23	5.88	6.63	6.39	5.47	3.85	2.20	1.86
Tilt adjusted each month	2.61 30°	3.40 38°	4.02 46°	4.64 54°	5.23 62°	6.22 70°	6.63 62°	6.42 54°	5.53 46°	4.14 38°	2.49 30°	2.21 22°

Pennsylvania

	Jan	Feb	Mar	Apr	May	Jun	Jul	Aug	Sep	Oct	Nov	Dec
Flat - 90°	1.87	2.65	3.52	4.32	5.11	5.44	5.50	4.86	4.03	3.07	1.97	1.58
Upright - 0°	2.91	3.16	3.00	2.63	2.46	2.35	2.45	2.64	3.05	3.39	2.78	2.62
50° angle Year-round tilt	3.07	3.74	4.17	4.38	4.71	4.82	4.96	4.74	4.55	4.16	3.03	2.69
34° angle Best winter tilt	3.23	3.80	4.04	4.03	4.17	4.18	4.33	4.28	4.33	4.18	3.16	2.86
66° angle Best summer tilt	2.73	3.48	4.09	4.53	5.05	5.26	5.37	4.98	4.54	3.92	2.75	2.38
Tilt adjusted each month	3.23 34°	3.80 42°	4.17 50°	4.54 58°	5.05 66°	5.46 74°	5.37 66°	5.01 58°	4.55 50°	4.20 42°	3.16 34°	2.88 26°

Rhode Island

	Jan	Feb	Mar	Apr	May	Jun	Jul	Aug	Sep	Oct	Nov	Dec
Flat - 90°	1.89	2.69	3.75	4.54	5.31	5.76	5.57	5.07	4.13	3.07	1.98	1.62
Upright - 0°	3.25	3.44	3.40	2.85	2.60	2.52	2.55	2.83	3.26	3.58	3.04	2.99
48° angle Year-round tilt	3.33	3.97	4.61	4.67	4.93	5.04	5.04	5.01	4.76	4.31	3.24	2.98
32° angle Best winter tilt	3.53	4.06	4.49	4.30	4.35	4.36	4.40	4.52	4.55	4.34	3.39	3.19
64° angle Best summer tilt	2.93	3.67	4.49	4.82	5.28	5.49	5.45	5.25	4.74	4.03	2.91	2.60
Tilt adjusted each month	3.53 32°	4.06 40°	4.61 48°	4.82 56°	5.28 64°	5.77 72°	5.45 64°	5.28 56°	4.76 48°	4.36 40°	3.39 32°	3.23 24°

South Carolina

	Jan	Feb	Mar	Apr	May	Jun	Jul	Aug	Sep	Oct	Nov	Dec
Flat - 90°	2.52	3.14	4.28	5.41	5.88	5.90	5.75	5.13	4.44	3.80	2.83	2.39
Upright - 0°	3.33	3.19	3.26	2.85	2.37	2.18	2.23	2.46	2.93	3.67	3.59	3.49
56° angle Year-round tilt	3.67	4.03	4.89	5.47	5.40	5.23	5.18	4.95	4.81	4.80	4.05	3.67
40° angle Best winter tilt	3.87	4.09	4.76	5.05	4.77	4.53	4.53	4.49	4.58	4.83	4.25	3.93
72° angle Best summer tilt	3.27	3.76	4.77	5.62	5.79	5.70	5.60	5.19	4.79	4.50	3.64	3.22
Tilt adjusted each month	3.87 40°	4.09 48°	4.89 56°	5.62 64°	5.79 72°	5.90 80°	5.60 72°	5.21 64°	4.81 56°	4.85 48°	4.25 40°	3.97 32°

South Dakota

	Jan	Feb	Mar	Apr	May	Jun	Jul	Aug	Sep	Oct	Nov	Dec
Flat - 90°	1.75	2.54	3.61	4.79	5.83	6.51	6.60	5.75	4.42	3.04	1.93	1.47
Upright - 0°	3.44	3.52	3.48	3.16	2.91	2.84	2.99	3.33	3.77	3.91	3.40	3.04
46° angle Year-round tilt	3.45	3.98	4.58	5.02	5.41	5.62	5.85	5.76	5.33	4.57	3.53	3.01
30° angle Best winter tilt	3.68	4.07	4.45	4.60	4.73	4.80	5.03	5.17	5.08	4.62	3.71	3.21
62° angle Best summer tilt	3.03	3.67	4.47	5.19	5.83	6.19	6.39	6.06	5.28	4.27	3.14	2.62
Tilt adjusted each month	3.68 30°	4.07 38°	4.58 46°	5.19 54°	5.83 62°	6.54 70°	6.39 62°	6.09 54°	5.33 46°	4.63 38°	3.71 30°	3.25 22°

Tennessee

	Jan	Feb	Mar	Apr	May	Jun	Jul	Aug	Sep	Oct	Nov	Dec
Flat - 90°	2.06	2.72	3.83	4.97	5.47	5.79	5.79	5.30	4.63	3.62	2.43	1.88
Upright - 0°	2.72	2.81	3.00	2.77	2.38	2.26	2.33	2.63	3.23	3.69	3.13	2.69
54° angle Year-round tilt	3.00	3.51	4.38	5.02	5.04	5.13	5.22	5.14	5.13	4.70	3.52	2.87
38° angle Best winter tilt	3.14	3.54	4.25	4.63	4.46	4.45	4.55	4.65	4.90	4.74	3.67	3.03
70° angle Best summer tilt	2.71	3.29	4.30	5.17	5.40	5.58	5.64	5.39	5.10	4.41	3.18	2.55
Tilt adjusted each month	3.14 38°	3.55 46°	4.38 54°	5.18 62°	5.40 70°	5.79 78°	5.64 70°	5.42 62°	5.13 54°	4.76 46°	3.67 38°	3.05 30°

Texas

	Jan	Feb	Mar	Apr	May	Jun	Jul	Aug	Sep	Oct	Nov	Dec
Flat - 90°	2.83	3.41	4.40	5.25	5.62	6.36	6.56	5.94	4.97	4.05	3.07	2.63
Upright - 0°	3.35	3.21	3.07	2.57	2.16	2.05	2.14	2.50	3.03	3.55	3.45	3.35
60° angle Year-round tilt	3.82	4.18	4.86	5.27	5.19	5.65	5.90	5.72	5.31	4.87	4.07	3.67
44° angle Best winter tilt	4.03	4.25	4.74	4.89	4.61	4.88	5.12	5.18	5.08	4.90	4.25	3.91
76° angle Best summer tilt	3.41	3.89	4.75	5.39	5.54	6.15	6.38	5.98	5.27	4.57	3.66	3.24
Tilt adjusted each month	4.03 44°	4.25 52°	4.86 60°	5.39 68°	5.54 76°	6.36 84°	6.38 76°	6.00 68°	5.31 60°	4.92 52°	4.25 44°	3.95 36°

Utah

	Jan	Feb	Mar	Apr	May	Jun	Jul	Aug	Sep	Oct	Nov	Dec
Flat - 90°	2.31	3.11	4.47	5.54	6.55	7.31	7.08	6.11	5.10	3.76	2.45	2.07
Upright - 0°	4.10	3.99	4.10	3.36	2.91	2.73	2.83	3.21	4.03	4.53	3.92	4.07
50° angle Year-round tilt	4.17	4.62	5.61	5.76	6.04	6.38	6.33	6.04	6.01	5.43	4.14	3.99
34° angle Best winter tilt	4.46	4.74	5.49	5.30	5.27	5.41	5.43	5.43	5.76	5.52	4.37	4.32
66° angle Best summer tilt	3.64	4.24	5.42	5.93	6.51	7.03	6.91	6.34	5.93	5.03	3.67	3.43
Tilt adjusted each month	4.46 34°	4.74 42°	5.61 50°	5.93 58°	6.51 66°	7.34 74°	6.91 66°	6.37 58°	6.01 50°	5.53 42°	4.37 34°	4.39 26°

Vermont

	Jan	Feb	Mar	Apr	May	Jun	Jul	Aug	Sep	Oct	Nov	Dec
Flat - 90°	1.61	2.53	3.54	4.34	4.86	5.43	5.49	4.81	3.72	2.45	1.50	1.26
Upright - 0°	3.06	3.55	3.42	2.87	2.53	2.54	2.65	2.84	3.06	2.85	2.30	2.38
46° angle Year-round tilt	3.08	4.00	4.49	4.51	4.50	4.72	4.89	4.77	4.33	3.41	2.46	2.40
30° angle Best winter tilt	3.27	4.09	4.37	4.14	3.97	4.08	4.26	4.30	4.11	3.41	2.55	2.54
62° angle Best summer tilt	2.72	3.68	4.38	4.66	4.84	5.17	5.31	5.02	4.32	3.24	2.24	2.11
Tilt adjusted each month	3.27 30°	4.09 38°	4.49 46°	4.67 54°	4.84 62°	5.45 70°	5.31 62°	5.05 54°	4.33 46°	3.43 38°	2.55 30°	2.56 22°

Virginia

	Jan	Feb	Mar	Apr	May	Jun	Jul	Aug	Sep	Oct	Nov	Dec
Flat - 90°	2.17	2.83	3.89	4.76	5.38	5.63	5.56	4.98	4.28	3.58	2.43	1.94
Upright - 0°	3.11	3.10	3.15	2.73	2.42	2.28	2.34	2.56	3.04	3.79	3.32	3.00
53° angle Year-round tilt	3.34	3.79	4.53	4.82	4.97	5.00	5.02	4.83	4.74	4.75	3.65	3.12
37° angle Best winter tilt	3.52	3.84	4.40	4.44	4.39	4.34	4.39	4.38	4.52	4.79	3.82	3.32
69° angle Best summer tilt	2.98	3.53	4.43	4.97	5.31	5.44	5.42	5.07	4.72	4.44	3.28	2.74
Tilt adjusted each month	3.52 37°	3.85 45°	4.53 53°	4.97 61°	5.31 69°	5.64 76°	5.42 69°	5.09 61°	4.74 53°	4.81 45°	3.82 37°	3.35 30°

Washington

	Jan	Feb	Mar	Apr	May	Jun	Jul	Aug	Sep	Oct	Nov	Dec
Flat - 90°	1.06	1.93	2.94	4.05	4.99	5.51	5.88	5.20	3.98	2.25	1.25	0.91
Upright - 0°	1.55	2.42	2.69	2.76	2.73	2.72	2.97	3.23	3.42	2.62	1.79	1.43
43° angle Year-round tilt	1.68	2.79	3.52	4.15	4.55	4.76	5.21	5.18	4.68	3.13	1.95	1.51
27° angle Best winter tilt	1.72	2.79	3.37	3.80	4.01	4.10	4.51	4.65	4.43	3.10	1.99	1.56
59° angle Best summer tilt	1.56	2.65	3.50	4.33	4.91	5.24	5.68	5.47	4.70	3.00	1.82	1.39
Tilt adjusted each month	1.72 27°	2.81 35°	3.532 43°	4.34 51°	4.91 59°	5.55 66°	5.68 59°	5.51 51°	4.68 43°	3.14 35°	1.99 27°	1.56 20°

West Virginia

	Jan	Feb	Mar	Apr	May	Jun	Jul	Aug	Sep	Oct	Nov	Dec
Flat - 90°	1.85	2.49	3.51	4.58	5.09	5.82	5.63	4.99	4.31	3.33	2.12	1.65
Upright - 0°	2.56	2.66	2.84	2.68	2.37	2.36	2.39	2.61	3.14	3.53	2.81	2.46
52° angle Year-round tilt	2.79	3.27	4.05	4.64	4.70	5.15	5.07	4.85	4.82	4.42	3.13	2.59
36° angle Best winter tilt	2.91	3.29	3.92	4.27	4.16	4.45	4.43	4.39	4.59	4.44	3.25	2.74
68° angle Best summer tilt	2.51	3.07	3.98	4.79	5.03	5.62	5.49	5.09	4.80	4.15	2.84	2.31
Tilt adjusted each month	2.91 36°	3.30 44°	4.05 52°	4.79 60°	5.03 68°	5.83 76°	5.49 68°	5.12 60°	4.82 52°	4.46 44°	3.25 36°	2.75 28°

Wisconsin

	Jan	Feb	Mar	Apr	May	Jun	Jul	Aug	Sep	Oct	Nov	Dec
Flat - 90°	1.80	2.67	3.58	4.49	5.33	5.97	5.87	5.05	4.01	2.76	1.79	1.52
Upright - 0°	3.34	3.63	3.34	2.89	2.67	2.63	2.71	2.89	3.25	3.23	2.82	3.05
47° angle Year-round tilt	3.39	4.13	4.45	4.64	4.92	5.18	5.24	4.99	4.67	3.87	3.00	3.01
31° angle Best winter tilt	3.60	4.22	4.32	4.25	4.33	4.45	4.54	4.49	4.44	3.88	3.12	3.22
63° angle Best summer tilt	2.98	3.81	4.35	4.80	5.30	5.67	5.71	5.25	4.66	3.65	2.70	2.61
Tilt adjusted each month	3.60 31°	4.22 39°	4.45 47°	4.80 55°	5.30 63°	5.99 70°	5.71 63°	5.29 55°	4.67 47°	3.90 39°	3.12 31°	3.26 24°

Wyoming

	Jan	Feb	Mar	Apr	May	Jun	Jul	Aug	Sep	Oct	Nov	Dec
Flat - 90°	2.13	2.92	4.10	5.13	6.14	6.92	6.65	5.82	4.79	3.52	2.34	1.89
Upright - 0°	3.81	3.78	3.76	3.18	2.84	2.73	2.80	3.15	3.83	4.28	3.83	3.76
49° angle Year-round tilt	3.88	4.37	5.12	5.32	5.66	5.98	5.96	5.76	5.65	5.11	4.03	3.69
33° angle Best winter tilt	4.13	4.47	4.99	4.89	4.96	5.09	5.13	5.18	5.40	5.18	4.25	3.99
65° angle Best summer tilt	3.39	4.02	4.97	5.49	6.10	6.58	6.50	6.05	5.59	4.75	3.58	3.18
Tilt adjusted each month	4.13 33°	4.47 41°	5.12 49°	5.49 57°	6.10 65°	6.93 72°	6.50 65°	6.08 57°	5.65 49°	5.19 41°	4.25 33°	4.05 26°

Appendix C – Typical Power Requirements

When creating your power analysis, you need to establish the power requirements for your system. The best way is to measure the actual power consumption using a watt meter.

Finding a ballpark figure for similar devices is the least accurate way of finding out your true power requirements. However, for an initial project analysis it can be a useful way of getting some information quickly:

Household and office

Device	Wh
Air conditioning	2500
Air cooling	700
Cell phone charger	10
Central heating pump	800
Central heating controller	20
Clothes dryer	2750
Coffee maker – espresso	1200
Coffee percolator	600
Computer systems:	
– Broadband modem	25
– Broadband and wireless	50
– Desktop PC	240
Document scanner	40
– Laptop	45
– Monitor – 17" flat screen	70
– Monitor – 19" flat screen	85
– Monitor – 22" flat screen	120
– Netbook	15
– Network hub – large	100
– Network hub – small	20
– Inkjet printer	250
– Laser printer	350
– Server – large	2200
– Server – small	1200
Deep fat fryer	1450
Dishwasher	1200
Electric blanket – double	100

Device	Wh
Electric blanket – single	50
Electric cooker	10000
Electric toothbrush	1
Fan – ceiling	80
Fan – desk	60
Fish tank	5
Food mixer	130
Fridge – 12 cu. ft.	280
Fridge – caravan fridge	110
Fridge – solar energy saving	5
Fridge-freezer – 16 cu. ft.	350
Fridge-freezer – 20 cu. ft.	420
Hair dryer	1000
Heater – fan	2000
Heater – halogen spot heater	1000
Heater – oil filled radiator	1000
Heater – underfloor (per m²)	80
Iron	1000
Iron – steam	1500
Iron – travel	600
Kettle	2000
Kettle – travel	700
Lightbulb – energy saving	11
Lightbulb – fluorescent	60
Lightbulb – halogen	50
Lightbulb – incandescent	60
Microwave oven – large	1400

Device	Wh		Device	Wh
Microwave oven – small	900		– LCD 20"	80
Music system – large	250		– LCD 24"	120
Music system – small	80		**Device**	**Wh**
Photocopier	1600		– LCD 32"	200
Device	**Wh**		– DVD player	80
Power shower	240		– Set top box	25
Radio	15		– Video games console	45
Sewing machine	75		Toaster	1200
Shaver	15		Upright freezer	250
Slow cooker	200		Vacuum cleaner	700
Television:			Washing machine	550
– LCD 15"	50		Water heater – immersion	1000

Garden and DIY

Device	Wh		Device	Wh
Concrete mixer	1400		– Hover mower – large	1400
Drill:			Lawn raker	400
– Bench drill	1500		Pond:	
– Hammer drill	1150		– Small filter	20
– Handheld drill	700		– Large filter	80
– Cordless drill charger	100		– Small fountain pump	50
Electric bike charger	100		– Large fountain pump	200
Flood light:			Rotavator	750
– Halogen – large	500		Saw:	
– Halogen – small	150		– Chainsaw	1150
– Fluorescent	36		– Jigsaw	550
– LED	1		– Angle grinder – small	1050
Hedge trimmer	500		– Angle grinder – large	2000
Lathe – small	650		Shed light:	
Lathe – large	900		– Large energy saving	11
Lawn mower:			– Small energy saving	5
– Cylinder mower – small	400		Strimmer – small	250
– Cylinder mower – large	700		Strimmer – large	500
– Hover mower – small	900			

Caravans, boats and recreational vehicles

Device	Wh		Device	Wh
Air cooling	400		– Cool box – small	50
Air heating	750		– Cool box – large	120
Coffee percolator	400		– Electric/gas fridge	110
Fridge:			– Low energy solar fridge	5

Kettle	700	Halogen lighting	10
Fluorescent light	10	LED lighting	1

Appendix D – Living Off-Grid

Living off-grid is an aspiration for many people. You may want to 'grow your own' electricity and not be reliant on electricity companies. You may live in the middle of nowhere and be unable to get an outside electricity supply. Whatever your motive, there are many attractions for using solar power to create complete self-sufficiency.

Do not confuse living off-grid with a grid-tie installation and achieving a balance where energy exported to the grid minus energy imported from the grid equals a zero overall import of electricity. A genuinely off-grid system means you use the electricity you generate every time you switch on a light bulb or turn on the TV. If you do not have enough electricity, nothing happens.

Before you start, be under no illusions. This is going to be an expensive project and for most people it will involve making some significant compromises on power usage in order to make living off-grid a reality.

In this book, I have been using the example of a holiday home. The difference between a holiday home and a main home is significant: if you are planning to live off-grid all the time, you may not be so willing to give up some of the creature comforts that this entails. Compromises that you may be prepared to accept for a few days or weeks may not be so desirable for a home you are living in for fifty-two weeks a year.

Remember that a solar electric system is a long-term investment, but will require long-term compromises as well. You will not have limitless electricity available when you have a solar electric system, and this can mean limiting your choices later on. If you have children at home, consider their needs as well: they will increase as they become teenagers and they may not be so happy about making the same compromises that you are.

You need to be able to provide enough power to live through the winter as well as the summer. You will probably use more electricity during the winter than the summer: more lighting and more time spent inside the house mean higher power requirements.

Most off-grid installations involve a variety of power sources: a solar electric system, a wind turbine and possibly a hydro-generation system if you have a fast-flowing stream with a steep enough drop. Of these technologies, only hydro on a suitable stream has the ability to generate electricity 24 hours a day, seven days a week.

In addition to using solar, wind and hydro for electrical generation, a solar water system will help heat up water and a ground source heat pump may be used to help heat the home.

When installing these systems in a home, it is important to have a *failover* system in place. A failover is simply a power backup so that if the power generation is insufficient to cope with your needs, a backup system cuts in.

Diesel generators are often used for this purpose. Some of the more expensive solar controllers have the facility to work with a diesel generator, automatically starting up the generator in order to charge up your batteries if the battery bank runs too low on power.

Advanced solar controllers with this facility can link this in with a timer to make sure the generator does not start running at night when the noise may be inappropriate.

A solar electric system in conjunction with grid electricity

Traditionally, it has rarely made economic sense to install a solar electric system for this purpose. This has changed over the past three years, with the availability of financial assistance in many parts of the world.

If you are considering installing a system purely on environmental grounds, make sure that what you are installing actually does make a difference to the environment. If you are planning to sell back electricity to the utility grids during the day, then unless peak demand for electricity in your area coincides with the times your solar system is generating electricity, you are actually unlikely to be making any real difference whatsoever.

A solar energy system in the southern states of America can make a difference to the environment, as peak demand for electricity tends to be when the sun is shining and everyone is running air conditioning units. A grid-tie solar energy system in the United Kingdom is unlikely to make a real difference to the environment unless you are using the electricity yourself or you live in an industrial area where there is high demand for electricity during the day.

If you are in the United Kingdom or Canada and are installing a solar energy system for the primary motive of reducing your carbon impact, a grid fallback system is the most environmentally friendly solution. In this scenario, you do not export energy back to the grid, but store it and use it yourself. When the batteries have run down, your power supply switches back to the grid. There is more information on grid fallback in Appendix E.

There may be other factors that make solar energy useful. For example, ensuring an electrical supply in an area with frequent power cuts, using the solar system in conjunction with an electric car, or for environmental reasons where the environmental benefits of the system have been properly assessed.

One of the benefits of building a system to work in conjunction with a conventional power supply is that you can take it step by step, implementing a smaller system and growing it as and when finances allow.

As outlined in Chapter Three, there are three ways to build a solar electric system in conjunction with the grid: a grid-tie system, a grid interactive system and grid fallback.

You can choose to link your solar array into the grid as a grid-tied system if you wish, so that you supply electricity to the grid when your solar array is generating the majority of its electricity and you use the grid as your battery. It is worth noting that if there is a power cut in your area, your solar electric system will be switched off as a safety precaution, which means you will not be able to use the power from your solar electric system to run your home, should there be a power cut.

Alternatively, you can design a stand-alone solar electric system to run some of your circuits in your house, either at grid-level AC voltage or on a DC low-voltage system. Lighting is a popular circuit to choose, as it is a relatively low demand circuit to start with.

As a third alternative, you can wire your solar electric system to run some or all of the circuits in your house, but use an AC relay to switch between your solar electric system when power is available, and electricity from the grid when your battery levels drop too low. In other words, you are using the grid as a power backup, should your solar electric system not provide enough power. This setup is known as a grid fallback system. A diagram showing this configuration is shown in the next Appendix under the section on grid fallback.

Appendix E – Other Solar Projects

Grid fallback system/ grid failover system

Grid fallback and grid failover are both often overlooked as a configuration for solar power. Both these systems provide AC power to a building alongside the normal electricity supply and provide the benefit of continued power availability in the case of a power cut.

For smaller systems, a solar electric emergency power system can be cost-competitive with installing an emergency power generator and uninterruptable power supplies. A solar electric emergency power system also has the benefit of providing power all of the time, thereby reducing ongoing electricity bills as well as providing power backup.

The difference between a grid fallback system and a grid failover system is in the configuration of the system. A grid fallback system provides solar power for as much of the time as possible, only switching back to the grid when the batteries are flat. A grid failover system cuts in when there is a power cut.

Most backup power systems provide limited power to help tide premises over a short-term power cut of 24 hours or less. Typically, a backup power system would provide lighting, enough electricity to run a heating system and enough electricity for a few essential devices.

As with all other solar projects, you must start with a project scope. An example scope for a backup power project in a small business could be to provide electricity for lighting and for four PCs and to run the gas central heating for a maximum of one day in the event of a power failure.

If your premises have a number of appliances that have a high-energy use, such as open fridges and freezer units, for example, it is probably not cost-effective to use solar power for a backup power source.

Installing any backup power system will require a certain amount of rewiring. Typically, you will install a secondary distribution panel (also known as a consumer unit) containing the essential circuits, and connect this after your main distribution panel. You then install an AC relay or a transfer switch between your main distribution panel and the secondary distribution panel, allowing you to switch between your main power source and your backup source:

In this above diagram, a second consumer box has been wired into the electrical system, with power feeds from both the main consumer box and an inverter connected to a solar system.

Switching between the two power feeds is an automatic transfer box. If you are configuring this system to be a grid fallback system, this transfer box is configured to take power from the solar system when it is available, but then switches back to grid-sourced electricity if the batteries on the solar system have run down.

This provides a backup for critical power when the normal electricity supply is not available, but also uses the power from the solar system to run your devices when this is available.

If you are configuring this system to be a grid failover system, the transfer box is configured to take power from the normal electricity supply when it is available, but then switches to power from the solar system if it is not.

One issue with this system is that when the transfer box switches between one power source and the other, there may be a very short loss of power of around $^1/_{20}$ of a second. This will cause lights to flicker momentarily, and in some cases may reset electronic equipment such as computers, TVs and DVD players.

Many modern transfer boxes transfer power so quickly that this is not a problem. However, if you do experience this problem it can be resolved by installing a small uninterruptable power supply (UPS) on any equipment affected in this way.

You can buy fully built-up automatic transfer boxes, or you can build your own relatively easily and cheaply using a high-voltage AC Double-Pole/ Double-Throw (DPDT) Power Relay, wired so that when the inverter is providing power, the relay takes power from the solar system, and when the inverter switches off, the relay switches the power supply back to the normal electricity supply.

Portable solar power unit

A popular and simple project, a briefcase-sized portable power unit allows you to take electricity with you wherever you go. They are popular with people who go camping, or with repair people who need to take small power tools to locations where they cannot always get access to electrical power.

In essence, a portable power unit comprises four components: a small solar panel, a solar controller, a sealed lead acid gel battery and an inverter, all built into a briefcase.

Many people who build them add extra bits as well. A couple of light bulbs are a popular addition, as is a cigarette lighter adapter to run 12-volt car accessories.

For safety purposes, it is important to use a sealed lead acid gel battery for this application, so that you can place the unit on its side, if necessary, without it leaking.

For occasional use, a portable solar power unit can be a good alternative to a petrol generator: they are silent in operation and extremely easy to use. Their disadvantage is that, once the battery is flat, you cannot use it until it is fully charged up again and on solar power alone this can take several days.

For this reason, solar powered generators often include an external charger so they can be charged up quickly when necessary.

Solar boat

Boating on inland waterways has been undergoing a revival in recent years, especially with small craft powered with outboard motors.

Electric outboard motors are also becoming extremely popular: they are lighter, more compact, easier to use and cheaper to buy than the equivalent petrol outboard motor. Best of all, their silent running and lack of vibration make them ideally suited to exploring inland waterways without disturbing the wildlife.

For a small open boat, a 100-watt electric motor will power the boat effectively. Depending on what they are made from, small, lightweight boats can be exceptionally light   a 5m (15 foot) boat may weigh as little as 20kg (44 pounds), whilst a simple 'cabin cruiser' constructed from alloy may weigh as little as 80kg (175 pounds). Consequently, they do not require a lot of power to provide ample performance.

An 80 amp-hour leisure battery will provide around eight hours of constant motor use before running flat. This is more than enough for most leisure activity. Because most boats are typically only used at weekends during the summer, a solar panel can be a good alternative to lugging around a heavy battery (the battery can quite easily weigh more than the rest of the boat!).

Provided the boat is moored in an area where it will capture direct sunlight, a 50–60 watt panel is normally sufficient to charge up the batteries over a period of a week, without any external power source.

Solar shed light

There are several off-the-shelf packages available for installing solar shed lights and these often offer excellent value for money when bought as a kit rather than buying the individual components separately.

However, the manufacturers of these kits tend to state the best possible performance of their systems based on optimum conditions. Consequently, many people are disappointed when the 'four hours daily usage' turns out to be closer to twenty minutes in the middle of winter.

Of course, if you have done a proper site survey and design, you will have identified this problem before you bought the system. If you need longer usage, you can then buy a second solar panel when you buy the kit in order to provide enough solar energy.

Solar electric bikes

Electric bikes and motorbikes are gaining in popularity and are an excellent way of getting around on shorter journeys.

Electric bikes with pedals and a top power-assisted speed of 15mph are road-legal across Europe, Australia, Canada and the United States. You can ride an electric bike from the age of fourteen.

Legally, they are regarded as normal bicycles and do not require tax or insurance. They typically have 200-watt or 250-watt motors (up to 750-watt motors are legal in North America). Most electric bikes have removable battery packs so they can be charged up off the bike and usually have a total capacity of 330–400 watt-hours and a range of 12–24 miles (20–40km).

Thanks to their relatively small battery packs, a number of owners have built a solar array that fits onto a garage or shed roof to charge up their bike batteries. This is especially useful when you have two battery packs. One can be left on charge while the second is in use on the bike.

A number of people have also fitted solar panels onto electric trikes in order to power the trike while it is on the move. Depending on the size of trike and the space available, it is usually possible to fit up to around 100 watts of solar panels to a trike, whilst some of the load-carrying trikes and rickshaws have enough space for around 200 watts of solar panels. Such a system would provide enough power to drive 15–20 miles during the winter, and potentially an almost unlimited range during the summer, making them a very practical and environmentally friendly form of personal transport.

Whilst the solar-only range may not seem that great, there are many drivers who live in a sunny climate and only use their cars for short journeys a few times each week. For these people, it could mean that almost all their driving could be powered from the sun.

Even in colder climates such as the United Kingdom and Northern Canada, solar power has its uses in extending the range of these cars: by trickle-charging the batteries during the daytime, the batteries maintain their optimum temperature, thereby ensuring a good range even in cold conditions.

Meanwhile, a number of electric car owners have already made their cars solar powered by charging up their cars from a larger home-based solar array, providing truly green motoring for much greater distances. Several electric car clubs have built very small and lightweight solar powered electric cars and tricycles and at least one electric car owners' club is planning to provide a solar roof to fit to existing electric cars in the coming year.

Appendix F – Building Your Own Solar Panels (and Why You Shouldn't)

A number of people have asked me about building their own solar panels from individual solar cells and asked for my opinion on a number of websites that make claims that you can build enough solar panels to power your home for around $200 (£120).

I have a huge amount of respect for people with the aptitude and the ability to build their own equipment. These people often derive a great deal of personal satisfaction from being able to say, "I built that myself." Largely, these people are to be encouraged. If you want to build your own solar panels, however, I would advise caution.

There have been many claims made from certain websites that say it is possible to build your own solar panels and run your entire house on solar for an outlay of $200 or less, selling excess power back to the utility grid and even generating an income from solar.

Most of the claims made by these websites are either false or misleading. When you subscribe to these services, you typically receive the following:

- Instructions on how to build a solar panel that are virtually identical to instructions that are available free from sites like *instructables.com*
- Information on tax credits and rebates for installing solar PV in the United States. (However, these credits and rebates are not applicable for home-built equipment. The websites omit to tell you that.)
- A list of companies and eBay sellers who will sell you individual solar cells

Many of the websites claim, or at least imply, that you can run your home on a solar panel built for around $200. In reality, this amount will buy you enough solar cells to build a solar panel producing 60–120 watts, which is certainly not enough to allow you to run your home on solar power.

Leaving aside the obvious point that you can buy a cheap, but professionally built 60–100 watt solar panel with five-year warranty and anticipated 25-year lifespan for around $200 (£120) if you shop around, there are various reasons why it is not a good idea to build your own solar panels using this information:

- A solar panel is a precision piece of equipment, designed to survive outside for decades of inclement weather and huge temperature variation including intense heat
- Professionally manufactured solar panels use specifically designed components. They are built in a clean room environment to very high standards. For example, the glass is a

special tempered product designed to withstand huge temperatures and ensure maximum light penetration with zero refraction

- The solar cells you can buy from sellers on eBay are factory seconds, rejected by the factory. Many of them are blemished or chipped and damaged. They are extremely fragile, almost as thin as paper, brittle like glass and very easy to break

- Unless you are an expert at soldering techniques, you are likely to create a cold solder joint between one or more solar cells. Cold solder joints inside a solar panel are likely to create a high temperature arc, which can start a fire

- There are several documented cases where home-made solar panels have caught fire and caused damage to people's homes. These fires are typically caused by poor quality soldering or the use of wrong materials

- Many of the websites promoting home-made solar panels claim that you can power your house with them. In the United States, connecting home-made panels to your household electrics would be in violation of the National Electric Code and you would therefore not be allowed a permit to install them

- Many of these websites imply that you can also sell your power back to the utility companies. It is actually illegal to install non-approved power generation equipment to the utility grid in many countries, including both the United States and the United Kingdom

- The tax credits and rebates that are available for installing solar PV on your home are not available for home-built solar panels

Many people who make their own solar panels have found that they fail after a few months due to moisture penetration, or fail after only a few days or weeks due to high temperature arcing and panel failure.

Most instructions recommend building a frame out of wood and covering it with Plexiglas or acrylic. This is extremely bad advice:

- Never build a solar panel frame and backing out of wood. This is dangerous because of the intense heat build-up in a solar panel. On a hot and sunny day, the surface temperature of the panel can exceed 90°C (175°F). If there is any additional heat build-up within the panel due to short circuits or poor quality soldering, these spot temperatures could be as high as 800°C (1,472°F). At these sorts of temperatures, you can easily start a fire

- Do not use Plexiglas or acrylic to cover your home-made solar panel. Tiny imperfections in the material can lead to light refractions and intense heat build-up on elements within the panel. Plexiglas and acrylic can also distort under high temperatures, increasing these light refractions over time. The effect can be like a magnifying glass, concentrating the intensity of the sunlight onto a small spot on the solar cell, which could result in fire

If you wish to build a small solar panel for fun, as a way of learning more about the technology, then you can get instructions on how to do this free of charge from many websites such as *instructables.com*. Build a small one as a fun project if you so wish. You will learn a lot about the technology by doing so. However:

- Treat your project as a learning exercise, not as a serious attempt to generate electricity
- Never build a solar panel with a wood frame
- Treat your home-made solar panel as a fire hazard
- Do not mount your completed home-made solar panel as a permanent fixture
- Only use your home-made solar panel under supervision, checking regularly for heat build-up on the solar panel or frame. Remember that the front of the solar panel may get extremely hot, especially on hot, sunny days. Do not touch the solar panel with your fingers
- Visually check your home-made solar panel every time you plug it in to ensure there is no moisture penetration. If you spot moisture penetration, stop using the solar panel immediately
- Use the cheapest solar charge controller you can find for your project. The warranty will be invalidated on the controller by using a home-made panel, but at least if you damage a cheap controller you haven't damaged an expensive one
- Never charge batteries using your home-made solar panel without using a solar charge controller
- Never run an inverter directly from your home-made solar panel

Also by Michael Boxwell:

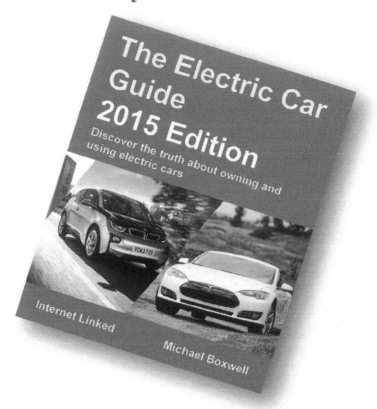

The 2015 Electric Car Guide is the most detailed and comprehensive book on electric car ownership available today. Written by best-selling author Michael Boxwell, who has been using electric cars as daily transport since 2006, these books are based on many years of practical experience in using battery powered cars on a day-to-day basis.

If you are new to electric cars, you've undoubtedly got lots of questions about them. Are they really environmentally friendly? What are they like to drive? Are they practical enough to be a realistic alternative? And what happens if they run out of range?

Superbly researched and clearly written, this book will explain exactly what it is like to own and use an electric car. Whether you have a particular model in mind, or if you are just interested in finding out more about them in general, The 2015 Electric Car Guide is written for you.

www.TheElectricCarGuide.net

Index

Made in the USA
Middletown, DE
29 March 2016